D0722423

The Innovative Choice

MARIO AMENDOLA AND JEAN-LUC GAFFARD

The Innovative Choice

An Economic Analysis of the
Dynamics of Technology

Basil Blackwell

First published 1988

Basil Blackwell Ltd
108 Cowley Road, Oxford, OX4 1JF, UK

Basil Blackwell Inc.
432 Park Avenue South, Suite 1503
New York, NY 10016, USA

British Library Cataloguing in Publication Data

Amendola, Mario
    The innovative choice: an economic
    analysis of the dynamics of technology.
    1. Technological innovations – Economic
    aspects
    I. Title   II. Gaffard, Jean-Luc
    338'.06   HC79.T4

    ISBN 0–631–15791–3

Library of Congress Cataloging in Publication Data

Amendola, Mario.
    The innovative choice.
    Bibliography: p.
    Includes index.
    1. Technological innovations – Economic aspects.
2. Economic development.   I. Gaffard, Jean-Luc.
II. Title.
HC79.T4A46 1988   338'.06   87–25004
ISBN 0–631–15791–3

Typeset in 10½ on 12 pt Times
by Dobbie Typesetting Service, Plymouth, Devon
Printed in Great Britain by T. J. Press, Padstow

Contents

Preface

One of the outstanding features of our time is represented by the extraordinary surge of innovations – which very often involve altogether new ways of matching economic and social needs and cause technology to become ever more deeply rooted in the environment. This points to the essentially economic nature of innovation, which is to be regarded as a pervasive process through which new forms of production and consumption are continuously brought into being.

The economic nature of the innovation process must be stressed at a time when the keenest academic minds have been concentrating rather on reinforcing the time-honoured distinction between the domain of economics and that of technology. It has thus been left to those who seek an understanding and explanation of the ongoing technological and productive transformations to turn more and more for support to sociology, epistemology, engineering and other disciplines.

In this book we attempt to fill the gap between the dominant trends in analysis and the problem of changes in technology by returning essentially to economic theory, as we are convinced that much of great relevance to the present issue is to be learnt from a theory which is well equipped to tackle the problem. In fact, as is usually the case in economics, what we really need is a change of viewpoint, so as to see the problems in their proper perspective, rather than new models and/or ever more sophisticated and finely tuned tools.

The change of perspective required consists, in our case, first of all in a shift of focus from the absorption into the prevailing productive structure of 'a given technological advance' conceived as a specific way to solve a certain problem (that is, a quantitative adjustment) to the process of 'creation of technology', interpreted as a process of modification of the environment regarded as a source of technology (that is, a qualitative phenomenon). In this perspective technology appears as the *result* of the process of innovation and not as the *pre-condition* for it. This makes it possible to go to the source of, and the actual inception of the changes, and not simply to deal with their effects. In this way, what most matters comes inside the process; the decision to start it off, the related sequence of choices that track its progress and the conditions for its viability all become relevant analytical issues.

This change of perspective, on the other hand, calls for an equally radical change in the analytical approach. It implies in fact a re-definition of some central concepts in Neo-Austrian terms, from the representation of the process of production to the interpretation of what has to be understood as capital. Furthermore, focusing on the process of innovation in itself, and interpretation of it as a qualitative change that gradually shapes new productive options and hence pushes a more or less predetermined point of arrival out of sight, requires a truly sequential analytical framework, in the Swedish tradition, in which money plays an essential role.

A process of this kind, we believe, is perfectly amenable to analysis with the tools of economic theory.

The Neo-Austrian theory recently proposed by J. R. Hicks provides a representation of the process of production that makes it possible to take explicitly into account the phase of construction of a new productive capacity as the physical expression of the specification of a new technology, and at the same time to consider the implications of the technical intertemporal complementarity of a process thus defined.

The classical concept of capital as a 'fund' – kept alive by Austrian and Swedish economists and essentially shared by Schumpeter, who pointed out its relevance when changes in the functioning of the economy are involved – is the natural complement to the Neo-Austrian representation of production in that it makes it possible to overcome the obstacle represented by the physical specification of capital goods when the object of the analysis is precisely that of the modifications in the forms of productive capacity.

When sequential learning is considered, as in the case of a process of creation of new productive options, money is essential. In fact, in this case, the process itself comes down to a related sequence of choices in a fundamentally uncertain context; also, in a sequence of this kind, the character of liquidity of money plays a central role. The consideration of this aspect of money goes back to Menger, long before Keynes developed his liquidity preference theory, and is certainly more effective than the latter when we come to the analysis of processes that take place in real, irreversible time.

The Swedish economists of the 1930s developed a sequence analysis that exhibits a process of change as a succession of periods, and that points explicitly to the role of expectations in determining how the successive periods actually link together in a sequence. This approach provides a general framework into which all the other elements can be integrated, thus laying the ground for the analysis of innovation as a process of qualitative change.

The problem is stated in chapter 1. The process of innovation is interpreted as a process of learning that brings about a modification of the environment regarded as a source of technology. This comes about as the result of an innovative choice, that is, of the decision to blaze new trails by starting processes of production whose features will only take on precise definitions along the way. The carrying out of innovative processes of production, in fact, implies research and experimentation, which help to specify the profiles of the processes themselves and at the same time cause the upgrading of the human resources that have been involved in those processes. Research and experimentation also lead to the appearance of entirely new skills and qualifications, which will themselves make it possible to devise and implement new forms of production and consumption. An innovative process is thus seen as a search for flexibility, in that it aims at extending the existing range of problems and solutions.

In chapter 2 we consider the ways in which economic theory has dealt with the problem of changes in technology and evaluate the capacity of the different approaches first to define the problem correctly and then to give it adequate treatment. We show, in particular, that the synchronic representation of production in models with a horizontal sectoral structure reduces technical change to a simple and analytically instantaneous reallocation of given inputs between the different sectors. We also show that circular relations in production are the main obstacle to the analysis of the process by which the productive capacity of the economy is transformed so as to acquire a different specification. The explicit consideration of a real time dimension, along which to articulate the process of change of productive capacity, implies first of all that capital goods need to be dealt with in a different way with respect to such models, and this, in turn, requires a different way of conceiving the process of production, with a return to the idea that 'production takes time.'

In chapter 3 we work out a 'constraints–decisions–constraints' sequence, which combines the analytical ingredients singled out in the previous chapters and provides the framework for the analysis of a process of change that takes place in a given environment, while involving modifications of the environment itself. The case for an innovative choice arises when there is a loss of confidence in what is going on, and it is portrayed as the breaking of a sequence. An increase in the demand for liquid assets then becomes the most immediate expression of the search for flexibility on the part of the economic agents who are faced with the appearance of uncertainty and hence with the prospect of learning. However, when learning is seen rightly as the result of doing, not of waiting, the appropriate result of this search for flexibility can only be an innovative choice: the central issue then becomes the viability of such a choice, which is strictly related to the determinants of the evolution of the process of innovation – in other words, to the whole set of environmental conditions and their changes. As we shall see, both the appearance of the problem of an innovative choice and the analysis of the ensuing process of innovation portrayed as a sequential process of learning call for the presence of money, which, for simplicity, we assume to be the only financial asset.

Chapter 4 may appear to be a break in the chain of thought followed in the previous chapters, but this is not really so. In fact, the model proposed is not one to be considered in the usual sense for its mathematical structure and for the analytical solutions that it is likely to admit, but rather as an attempt to verify the consistency and to explore the articulation in time of the sequence that portrays the process of innovation, in order to bring to light its relevant moments and their connections. In particular, the model makes it possible to analyse by means of non-stochastic simulations the patterns of evolution of the economy in various scenarios corresponding to different hypotheses as to the nature and the strength of the existing constraints, as to the behaviour of economic agents and as to the economic policy followed.

The results of the numerical experiments performed are analysed in chapter 5, in which we show how the proposed change of perspective provides an altogether different approach, not only to analysis, but also to policy implications. In this perspective, in fact, economic policy is no longer used for making a choice but aims at rendering an innovative choice viable, which is the only adequate reaction when the breaking of a sequence demands a qualitative change. In the last analysis, policy interventions essentially must affect the intensity and the pace of the learning process, both in production and in consumption, which are shown to be the main determinants of the viability of an innovative path. An economic policy conducive to change and innovation thus appears as an integrated set of active interventions, which must take into account the interactions over time of supply and demand and which must be considered not so much in terms of their quantitative aspects as for their articulation through time.

Maria Luisa Petit, Didier Austry, Sergio Bruno, Pierre Bernard and Daniel Gabay all helped in different ways and at different stages of this work, although they bear no responsibility for the way we availed ourselves of their help. Our greatest debt is certainly to John Hicks – not so much for the many references to his ideas that pervade the book as for the basic approach to economic problems, for that way of being economists and doing economic analysis that one of us learned directly from him and transmitted to the other.

The preface to *Capital and Time* closes with the following remarks: 'I am well aware, in concluding this book, that the task I set myself is by no means completed . . . one catches sight of further questions which need to be examined. I am sure, however, that there are plenty of people who will want to examine them; it is time I called them in.' Our hope is that we have succeeded in pushing on along the lines that Sir John indicated, and have not let him down.

M.A.
J.L.G.

1

The Process of Innovation

The 'Traditional' Approach: Adjustment to a Given Technology

The different interpretations of technology and of its changes that have
been cropping up in fairly recent years[1] can be classified, in the first
instance, according to where the focus of the analysis is.

Two main interpretations can be singled out in this perspective. On
one side we have what we might by now consider as the 'traditional'
approach, stemming from the theory of technical progress as it has
been developed within the context of the dominant production theory.
The focus, in this approach, is on the point of arrival of the process
of change – that is, on the configuration of the productive capacity
of the economy (of the firm) that results from the adoption of a given
technological advance and is uniquely determined by the characteristics
of the latter. The analysis consists in deducing the effects of the change
on the relevant magnitudes of the economy (productivity, employment
etc.) from a comparison of the features of its productive structure before
and after the change.

The analytic ingredients of this approach are: (1) a technological
advance that is assumed to be fully defined and developed at its first
appearance, and which generally consists in a new production process

---

[1] The interest in technical progress, within the field of modern economic thought,
is quite recent. 'One has only to go back to the 1930s to find a time when the importance
of technical progress did not seem at all obvious. Few economists would then have
given it the priority which now seems so natural. It was mentioned, but it took a fairly
subordinate place' (Hicks, 1974b, p. 21).

1

embodied in durable capital equipment (a new 'technique'); (2) a given productive structure into which the innovation must be absorbed and (3) a given point of arrival of the process, represented by a productive capacity fully shaped according to the new technique.

The basic reference is to a process of production that consists essentially in transforming given inputs into final output by means of physically defined productive equipment ('machines'). The relevant aspect of the productive problem is regarded as (the utilization of) the given equipment, while the relevant aspect of the technological problem is regarded as the embodiment of the idea, or of the improvement, in the equipment itself. The specific features of the 'machines' (which reflect the 'technique') are represented in turn by the way in which the given inputs are combined (the 'technical coefficients'), while technical progress (a change in the technique) is identified with a modification of these coefficients.

The techniques are therefore fully specified in terms of technical coefficients, have their own autonomous dimension and exist, so to speak, in their own right. They can be defined in a general and abstract way and hence considered as given in the sense that they can be freely appropriated provided the required inputs are available.

Different techniques can then be classified and compared on the basis of given criteria, and the problem of the choice of the technique can be structured as a typical maximization process in a context in which the choice set is given and the outcomes of the choices are known. The solution is obvious: once a technique has been defined as *superior* according to some criterion (e.g. a higher profit rate for a given wage rate) it is automatically selected, and the only problem for the economy is then to adjust its productive capacity to it. In some models (the majority), the adjustment is not even considered, and there is a jump directly to the description of the new shape of the productive capacity; in other (more recent ones) the adjustment is considered, mainly in the sense that there is an attempt to find the conditions required for convergence on the new position.

The process of innovation, in this context, is identified with the diffusion of innovation, that is, with the extent and the speed at which the economy proceeds to adopt a superior technique. However, all the attention is concentrated on the outcome of this process which is assumed *ex ante*; the reference to the process is thus essentially of a formal character.

Adoption can be general and instantaneous, as happens in the aggregate shifting production function model (Solow, 1957), where full reversibility of investments allows immediate availability of resources for a total reshaping of the productive capacity of the economy; or gradual, as is the case in 'vintage' models (Salter, 1960; Solow, 1962) where a certain irreversibility allows only a gradual accrual of resources to be invested in the new 'machines'. On the other hand, focus on inter-firm diffusion rather than on the economy as a whole brings to light the role of decision factors – namely demand factors – besides the simple availability of resources, in determining the extent and the speed of the diffusion/adoption.

In this approach, in any case, *it is the economy that is adjusted to the technology through the diffusion process.*[2]

The Evolutionary Approach: Innovation as a Process

In more recent times a 'new' approach has emerged from scattered but converging analytical work whose origins can be traced back to intuitions and isolated contributions made in the past, and from extensive research on patterns of technical change, diffusion of particular innovations and processes of industrial transformation.

The new interpretation focuses on two main objections that have been levelled against the 'traditional' approach: (1) that the technology is not something fully realized at its first appearance, to be considered as given and as something on which to draw more or less freely and (2) that its development does not take place '*in vitro*' but in a specific environment that contributes to the range, to the nature and to the direction of the improvements, and that itself gets changed as the process of development goes on.

This implies abandonment of the traditional neoclassical representation of technology as a given choice set where the outcome of each choice is known, and of its changes as a more or less mechanical expansion of the production sets, whether spontaneously or as a result of research and development. Classical theory already involved an evolutionary process, rather than rational choice between known alternatives.[3] On the other hand, Schumpeter (1934) had underlined

---

[2] The consideration of the role of the supply factors in the diffusion process, and consequently of important interactions between the growth of demand and the growth of productive potential, highlights a relation between diffusion and induced innovation (Metcalfe, 1981). A breakdown of the scheme according to which a given innovation is diffused within an unchanging adoption environment opens the way for a representation in which the innovation and environment change as diffusion proceeds.

[3] In particular Marx 'analogized an economic system to a biological entity which is ever changing instead of to a chemical reaction that is tending to an equilibrium . . . he however appeared to view technical advance as a change in the environment that capitalists welcomed but reacted to, rather than initiated' (Kamien and Schwartz, 1982, p. 6).

the intrinsically uncertain nature of the innovative activity, with innovators who do not even know the probability distribution of the likely outcome of their activity. Knight (1921) argued that when we are considering innovative changes it is not possible to calculate the right things to do, as only the subsequent events will reveal the appropriateness or otherwise of the choices made. More recently Nelson and Winter (1982) have considered firms' behaviour, and changes in this behaviour that are no longer maximization procedures in given contexts but are derived from a biological evolution model.

The 'new' interpretation of the process of innovation has emerged from the piling up of contributions that move along these lines. In this interpretation innovation is seen not so much as the absorption of a given technological result into the prevailing productive structure, but rather as the process through which a new technology is developed step by step from an initial impulse as a particular answer to given problems in a specific environment – and through which, at the same time, the productive structures of the economy acquire their new features as the technology builds up. This process has an essentially sequential character, in the sense that at each given moment different paths are open and different decisions can be taken leading to different alternative developments; but not all the conceivable developments are feasible, because of constraints and endowments that have gradually accumulated along the particular path that has led to the present state.

It follows immediately that the point of arrival of this process – that is, the shape of the productive capacity that reflects the new technology – can no longer be uniquely determined *a priori*. It will depend in fact on the path actually followed, which will determine how much of the potential technological content of the innovation will be brought to the surface and actually exploited, and in what way. Different points of arrival will then be possible, corresponding to different alternative paths followed. The focus of the analysis, although still concentrated on the point of arrival, also has its effects on the process that led to that point – which thus acquires an analytical relevance of its own.

The economy, in this context, *no longer adjusts passively to the technology but becomes the instrument for determining the extent, the nature and the articulation through time of the development of the technology.*

The Development of a Technology in a Given Environment

Reference to some typical contributions will help to single out the most relevant aspects of the 'new' approach.

The point of departure of the process of innovation has been defined in different ways, all sharing however the image of a process that brings to the surface something already contained in the point of departure itself in a nutshell. Thus Dosi (1982) speaks of the appearance of the new 'technological paradigm', regarded as a given pattern for the solution of selected technological problems that defines contextually the needs that one aims to fulfil, the scientific principles utilized for the task and the material technology to be used. Others refer to a change in a 'technological regime' (Nelson and Winter, 1977), or to a 'technological guidepost' (Sahal, 1981) – concepts that share with the paradigm the idea of a common approach to certain technological and productive problems. All these concepts can be taken to represent the original innovation, the impulse that sets off a process of development of a technology.

A 'technological trajectory' is then defined as the 'actualization of the promises contained in the new paradigm' (Dosi, 1982, p. 83), that is, the normal problem-solving activity leading from the scientific approach to the realized technology (-ies)[4] actually embodied in devices and equipment.[5]

The effective trajectory, that is, the path actually followed out of a much bigger set of possible ones, can be represented as 'the movement of multidimensional trade-offs between technological variables that the paradigm defines as relevant . . . and economic variables' (Dosi, 1982, p. 85) reflecting the environment in which the trajectory is traced out. New paradigms (new regimes, and so on) are in fact a necessary but not a sufficient condition for radically innovative choices leading to substantial technological and productive transformations: for the process of innovation to be actually feasible we also need the appropriate economic conditions. Thus the environment adds an economic dimension to the technological one in determining both the range of possible directions and the gradual specification of the path followed along one of these directions.

Such an analytical framework provides a convenient interpretation of the traditional distinction between major, or radical, innovations on one side and minor, or marginal, innovations on the other. Radical innovations can in fact be defined as extraordinary exogenous changes

[4] The emerging of a new paradigm often involves powerful complementarities between different technological developments and interdependencies between industries, thus giving rise to new 'technological systems' characterized by interrelated 'clusters' of technologies (like organic chemistry or semiconductor technologies) affecting different industries to varying degrees. See Freeman et al. (1982).

[5] The concept of a 'trajectory' was first introduced by Nelson and Winter (1977).

that bring about new technological paradigms (new regimes, and so on), and marginal, or incremental, innovations can be defined as endogenous changes concerning the normal progress along the trajectory that makes it possible to bring to light the potential content of a paradigm in a given environment. However, this distinction, on closer inspection, goes directly back to the original distinction between autonomous and induced inventions introduced by Hicks more than 50 years ago: 'We must put on one side those inventions which are the result of a change in the relative prices of the factors; let us call these *induced* inventions. The rest we may call *autonomous* inventions' (Hicks, 1932, p. 125). Hicks's definition thus points directly to the essence of the concept of marginal, or induced, innovations as opposed to the radical ones: that is, that the former are intrinsically linked to the process of development of a given technology, such as actually occurs within a specific environment, and are hence determined by economic factors reflecting the characteristics of this environment.[6]

The environment itself, however, changes together with the technology as the process of innovation goes on. This process, in fact, is at the same time one of development of the technology and of transformation of the productive structures of the economy, and it is characterized by a continuous feedback between technology and the environment. The appearance of a new paradigm (new regime, and so on) implies the breaking up of the existing industrial structure and a modification of the market conditions, followed by a gradual reshaping which reflects the scarcities encountered on the way, the particular effort made to overcome these scarcities, the changes in cost conditions, in profitability and in relative prices, the modifications of the consumers' preference system, and all the other events that represent the specific episodes that mark the actual profile of the process of innovation.

[6] More recently, Hicks (1974b) has offered an interpretation of the process of innovation based on the concept of 'Impulse', which is a blend of the original distinction between induced and autonomous inventions and his analysis of the 'traverse' (see chapter 2) as the transition to a superior technology. According to this interpretation, autonomous inventions express some kind of technological revolution and give an impulse to the economy. In the working out of the impulse other secondary inventions come about – 'the children of the original invention' – which represent the further developments and adaptations of the latter. The original impulse in fact is not inexhaustible; any expansion must sooner or later encounter scarcities, or run into bottlenecks, and some scarcities (such as those of land, or labour in general) cannot be removed in the same way as the shortages of some materials or of some kinds of skilled labour. Induced inventions are a way of overcoming such scarcities (mainly in that they permit substitution) and hence of keeping the 'Impulse' alive.

The Specific and Cumulative Character of Technical Change: Decreasing Returns and the Narrowing of the Options

The interpretation of technology and of its changes sketched out above highlights in particular the specific and cumulative character of the process of innovation.

If technology is not something on which it is possible to draw more or less freely provided the required resources are there, but is the result of an intentional effort in a given direction that goes on in time, then technological advances will not be random, but will be linked in a sequence, as they will be tailored to suit the specific interests, and to deal with the particular problems, of those who are actually engaged in an innovative activity and who are as a result acquiring growing experience while carrying this activity out.

The process of innovation will thus be necessarily characterized by cumulative effects at the level of the single firms and/or of the whole industry.[7] Cumulativeness, together with the specific character of the research and learning process, is on the other hand strictly related to private appropriability: it is mainly those who have obtained the improvements who can take full advantage of them. It has thus been suggested that an oligopolistic structure with increasing degrees of concentration is likely to develop on a technological trajectory, owing to the differential advantages of the first-comers, which, due precisely to cumulativeness, are not only stable through time, but might even tend to increase (Nelson and Winter, 1978; Dosi, 1984).

Technical change, then, is cumulative and specific; it builds upon existing skills and activities within firms, sectors and economic systems, and its trajectories vary according to the characteristics of the latter.[8] The process of innovation, it is worth repeating, is deeply rooted in the environment in which it takes place and that it helps to modify.

There is another aspect however – in addition to the strict connection between process and environment, but related to it – that should be

[7] 'Workers are learning to do their jobs better, management is learning how to organize more effectively, and engineers are redesigning the products to make the job easier and to replace labour where it is possible and economical to do so' (Nelson and Winter, 1982, p. 258).

[8] According to Pavitt (1984) trajectories of technological development in firms are strongly influenced by their principal activities, and can be classified and explained in terms of three main categories: supply-dominated, production-intensive and science-based firms.

underlined. A technological trajectory, whatever its character, direction and articulation through time, consists in developing a given performance potential already contained in the basic paradigm (new regime, and so on). However, 'any particular industrial technology has a finite scope for improvements in its technical performance. An undeveloped technology is akin to an exhaustible resource: as more is known less is left to discover' (Metcalfe and Gibbons, 1983, p. 7). The scope for improvements along a trajectory is thus limited, and further technological advances tend towards an asymptotic limit.

On the other hand, to the extent that these advances require the explicit commitment of specific and generic resources, scarcities will appear sooner or later whose effects cannot be postponed indefinitely by substitution processes and induced innovations, and that will therefore exert pressures resulting in rising costs of incremental advances as the limit of the trajectory is approached. Technological and economic factors thus interact to bring about decreasing returns as the process of innovation pushes on.

Nor is that all. As the technology gets more and more specified the direction of the trajectory becomes more and more uniquely determined,[9] restricting the range of the options left available and hence increasingly constraining the successive choices.[10] There is a point of arrival – which depends on the path actually followed but is already contained in the initial paradigm, although in an implicit form – that becomes clearer and clearer as the thrust along the trajectory gradually reduces the possible lines of development. This point of arrival, with its specific features, is no longer the only relevant moment of the analysis, as was the case in the 'traditional' approach; it is however always present on the scene although left in the shade in the first instant, when all the light is directed (but only instrumentally) on that process that in contrast the traditional approach left entirely in the dark.[11]

---

[9] Rosenberg (1969) speaks of 'technological imperatives' that guide the evolution of given technologies.

[10] As we pass from knowledge to artefact, illiquidities increase (particularly when strong economies of scale are involved), barring paths previously open at each successive step.

[11] The reference to a point of arrival that, although hidden, has been 'there' since the beginning, implies that the choices successively made along a trajectory have an essentially mechanical character. Nelson and Winter (1982, p. 68) speak of 'processes that involve a more or less mechanical following of a decision rule' rather than of 'processes that involve a considerable amount of deliberation'.

INNOVATION AS CREATION OF TECHNOLOGY

Generic Inputs and Specific Inputs in the Process of Production

We have pointed to an enlargement of the focus of the analysis – so as to include, together with the point of arrival of a change in the technology, the actual process that led to it – as to the main difference between the 'traditional' and the 'new' approach. Accordingly the process of innovation, a necessary and hence analytically irrelevant link between two points (or paths) in the one approach, has been seen to acquire a relevance of its own in the other.

It is now as well to underline the fact that the innovation, both when it is regarded as fully developed at its first appearance and when in contrast it is seen as acquiring a precise contour through a process, is generally conceived as the embodiment of some kind of advance (idea, approach, solution, . . .) in physical (productive) equipment. This is so because behind the interpretation of technology and of its changes, in both the approaches considered, there is the same image of the process of production: the traditional (and still explicitly or implicitly dominant) representation of a process which consists in the setting up and in the subsequent operation of a 'plant' made up of 'machines' which are often the sole meaningful expression of the technology selected.

In defining the phenomenon of production, the accent is put in this representation on the processing of generic (homogeneous or heterogeneous) inputs into finished goods and/or services with the aim of satisfying an external demand: the combination of these inputs defining the technology and, for short, the productive capacity (the 'machines') shaped according to it. *Generic* means here existing in their own right, outside and independently of the process of production in which they are employed, and hence available for use in different kinds of processes, characterized by different 'machines' embodying different 'techniques'.

This is immediately evident in the aggregate production function model, where different combinations of the homogeneous factors capital and labour stand directly for the different techniques; but the same holds for multi-sector models, where the processes in each sector are characterized by technical coefficients that reflect different combinations of the same (homogeneous or heterogenous) inputs, so that the reshaping of the productive capacity can be brought about by simply reallocating these inputs between the sectors (see chapter 2).

The relevant aspect of a process of production conceived as the transformation (through 'machines') of generic inputs into finished goods is then the *availability* of these inputs in given amounts, which will determine both the technique selected (the shape of the 'machines') and its actual adoption by the economy. This implies, in turn, an *analytically instantaneous* embodiment of the technique, in the sense that, once the productive capacity expression of such a technique has been defined in terms of a given combination of inputs, we are already dealing with the capacity itself from the analytic point of view, provided the required inputs are available in the right proportions.[12]

Quantitative Adjustment and Qualitative Change

What has just been said is true of the different approaches to changes in technology considered, both of which make reference to the above-mentioned representation of the process of production.

It is in fact the availability of generic resources that (a) is the main determinant of the more or less gradual process of diffusion/adoption when the process of innovation is interpreted as a simple 'adjustment to' a given technology and (b), when this process is seen instead as 'contributing to the development of' the technology, defines the economic trade-offs that actually characterize the technological trajectory, and brings about the induced inventions – directed to overcoming the scarcities arising during the process – that help to trace the direction of this trajectory.

Reference to the same view of the process of production, however, reduces the difference between the 'traditional' and the 'new' interpretation of the process of innovation to a simple matter of degree.

In both cases, in fact, this process becomes the expression of a 'quantitative adjustment'; although, in the second case, the fact of taking explicitly into account a development of the technology that might assume different profiles makes the same process appear as a 'qualitative change', in the sense of the creation of something new and different. However, in fact, nothing of the sort is contemplated *inside* the process itself. The qualitative change, from the analytic point of view, takes place *outside* the process of innovation, before the latter starts developing: it is 'the initial impulse', the 'appearance of a new technological paradigm', a 'change in the technological regime' or whatever we want to call it. The process of innovation in itself is just the quantitative development of something already essentially given: the gradual bringing to the surface of an initial 'technological content', the realization of a 'performance potential', or the like.

---

[12] This does not prevent us from considering a gradual accrual of these resources in time, as is the case with 'vintage' models.

Now, we maintain that the quantitative character of the analysis depends on the reference to a process of production that focuses on generic inputs whose most relevant analytical aspect is availability in larger or smaller amounts, and which assigns to the latter the role both of defining the (development of the) technology and of determining its actual adoption by the economy. That is why, in an attempt to go the whole way opened up by the 'new' approach and to define a process of innovation that contemplates (and makes it possible to analyse) a real qualitative change, we shall begin by making reference to a process of production implying the use of *specific* inputs: that is, inputs considered for the relevance of their characteristics in relation to the process itself and not simply for the amounts available.

Consideration of specific inputs is certainly not new in the literature, starting from the classics, who had seen the passage from craft production to factory production as marked by extreme specialization of the machines, which might even reach the point where they could have no employment alternative to their work in hand – in contrast to the corresponding changes for the labour force, which became increasingly less skilled as a result of the vertical subdivision of the process of production. In more recent times a return of interest in specific inputs after the long neoclassical interval – always with reference to the 'machines' – has characterized the analysis of the 'traverse' (Hicks, 1973; Lowe, 1976); this is an attempt to substitute the explicit consideration of the process of transformation of the productive structures of the economy for the logically instantaneous adjustment of the traditional analysis of technical progress.

Consideration of specific inputs, however, is not sufficient in itself to make it possible to deal with a qualitative change (see chapter 2, pp. 32–3). We need something more: we need to make reference to an altogether different image of the process of production, focusing on the central role played by specific resources, namely human resources – an image that appears more and more clearly through the mirror of the ongoing processes of industrial transformation.

The Changing Features of the Process of Production

It is a widely diffused and increasingly held opinion that we are confronted with such a surge of radical innovations that doubts are raised as to whether the existing analytical categories, schemes and classifications are still appropriate to describe the relevant aspects of the phenomenon of production as it takes place today, and to capture the even more radical changes that are expected to come about in the near future.

The emergence of modern science as the basis for an advancing technology is in fact bringing about a substantial modification of the characteristics – and, in some cases, of the very nature – of the phenomenon of production, implying quite often a new way of matching economic and social needs. Deep structural and organizational changes tend to obscure the very concepts of firm and factory, whose features and boundaries could once be precisely defined by physical elements (buildings, equipment, machines, . . .) but tend to disappear as 'hard' physical connections are replaced by 'soft' ones. This is associated, in many cases (consider activities such as maintenance, health, software and electronic systems, and engineering firms), with a tertiarization and a dematerialization of the process of production, resulting in what has been labelled as a transformation from the product industry to the function industry (Colombo, 1980), which tends to satisfy the market requirements by means of an integrated series of actions and interventions rather than by simply supplying commodities.

Besides these structural and organizational modifications, though, it is the very nature of the process of production itself that is likely to be affected. This is particularly true in the activities that are or that will be related to the technologies that are experiencing the most spectacular advances – such as microelectronics, which is already fully developing, and biotechnology, which envisages the application of biological organisms, systems and processes to manufacturing and service industries, and which over the next two decades is expected to affect a wide range of activities, from food and animal feed production to alternative energy sources, from waste recycling to medical and veterinary care.

In some highly specialized fields (reagents, diagnostics) the scale effect tends to disappear, along with the traditional subdivision of the process of production in the different phases (and of the firm in the different sectors) of research–development–production–distribution, and the like. Production in itself very often concerns in fact such small quantities that it is carried out by the research groups themselves along with the development of the know-how. In this kind of activity raw materials and, more generally, generic inputs play a less and less relevant role. Both the definition of the characteristics of the process of production and its effective articulation in a given context, and in time, take place while the process itself is being carried out; and this requires more and more specific resources (particular skills, specialized information etc.) which themselves get modified in the course of the process.

Furthermore, the integration of the phenomenon of production in the environment is becoming more and more important and is taking on widely different forms – and this is not just in the particular case of systems capable of offering not so much simple products as complex services that require suitable structures capable of receiving them (take for instance the current way of controlling parasites which through integrated biological intervention has replaced the simple and straightforward use of pesticides). Consulting, engineering and job production firms, and even manufacturing firms in the true classical sense, like those employed in the production of machines and instruments, are undergoing fundamental modifications in the production process which involve considerable amounts of upwards and downwards integration.

The heavy segmentation of the market and the fact that production in batches can practically be personalized as a result of the introduction of sophisticated electronic technology accentuates in particular the dependence on the characteristics of demand that tends to turn the customer into a participant in the production process to the point where he or she is no longer an external element but is a real specific input in this process. In fact the customer contributes more and more, not just to the detection of specific needs, but also to the definition of the problem of production, until a concrete solution is finally adopted and the specification of the particular characteristics of the product have been arrived at.

Against this changing background, the image of a process of production that consists in operating a pre-existing productive capacity to transform generic inputs into finished goods to meet the requirements of an external demand gives way to an activity of research and coordination – upstream of the strictly defined (and sometimes not even necessary) manufacturing phase – of the ingredients more suited to finding a solution to production problems (and hence to satisfying requirements) that often arise within the process itself. The essence of the process of production, therefore, is no longer embodied in devices and equipment, but lies in the characteristics of the specific inputs involved which themselves contribute to defining the profile of the process and its effective articulation contextually.

An essentially different type of production process, both from the technical and from the economic point of view, is thus emerging from the ongoing transformations; a process whose requirements and perspectives, as we shall see, cannot be met in the usual way by the existing industrial and economic structures. A clear definition of its features, however, is not all that important for the essential aspects of today's reality that it allows us to detect and to understand. More important is the change of perspective that it suggests in the interpretation of technology and of its changes. It is a change of perspective that will make it possible to throw light on the qualitative nature of the phenomenon of technical progress.

The Environment as a Source of Technology:
the Role of Human Resources

In comparing the above outline of the process of production with the traditional one, we have pointed out, as its most distinctive aspect, the effective articulation of the process in a given environment, reflecting the skills and the specific interests of those who are actually carrying it out.

In this perspective technology no longer appears as a specific way (with its own physical counterpart) of solving a given problem, but as an environment characterized by specific resources that make it possible to devise and implement different solutions to different problems. Accordingly, attention is shifted from the process of embodiment (or of development/embodiment) of a given technology to the process of modification of the environment as a source of technology.

The rooting of technology in the environment, on the other hand, implies that the specific character of the resources that are the expression of a given environment will be acquired through research, experimentation and learning processes that take place in the environment itself and that represent its particular history. These resources are then most easily interpreted as human resources, which are the natural depository of this history and whose specific character consists in particular skills that imply thinking out and organizing the productive problems in altogether original ways.[13] They must be looked at, in a wider sense, as referring both to the production and the consumption side, in a context in which the newly emergent forms of consumption also help to define new products and to articulate new processes of production.

In the same way, the process of innovation can no longer be thought of as 'the adjustment to' or even 'the development of' a given technology, but as a process of research and learning carried out in a given environment and resulting in the appearance of entirely new skills and qualifications that bring about a modification of the environment itself and thus make it possible to extend the existing range of problems and solutions.

From an analytic point of view, technology appears then as the *result* of the process of innovation, and not as a *pre-condition* of it. In this interpretation the process of innovation is seen as a process of 'creation of technology', that is, a process that makes it possible to deal analytically with the source of the changes and not only with their effects.

---

[13] The specific character of a resource, thus, does not depend on its particular nature, but on its relation with a given environment established through a process.

2

Changes in Technology
and Economic Theory

Intuition and Method in Classical Thought

Classical economists analysed the effects of changes in conditions of the economy by means of a comparison between positions corresponding to different conditions, labelled as *before* and *after* the change. The effects of changes in technology, in particular, were deduced from the comparison of states of the economy characterized by the use of different techniques: the adjustment of the productive capacity to the prevailing technique appearing as the preliminary requirement for the comparison to be possible.

The analytical implications of this method are quite obvious in the famous controversy on 'technological unemployment', and the related 'compensation' theory, associated with the problem of the introduction of machinery; a controversy that has divided the economists ever since (and before) the stage was set for it, in the chapter *On Machinery* which Ricardo added to the last edition of his *Principles* in 1821. As is well known, Ricardo argued in this chapter that the introduction of machinery could be detrimental to the interest of the working class, thus retracting his previous opinion that mechanization would be beneficial to all the different classes of society.

The first point to stress, with respect to this argument, is that the question of whether unemployment will result from the introduction of machinery, and what the reasons and the conditions are for this phenomenon to take place, is not the same as the question of whether

17

the displaced labour will be reabsorbed in the end, and what is required for this to occur. A first implication of the method of the comparison, in fact, is to mix up these problems, by focusing on the reabsorption of the labour that has been displaced by the introduction of machinery rather than on the phenomenon of technological unemployment in itself. The reason is that the problem of 'compensation' – which arises when the use of machinery is considered, that is, when machinery has already been built, set to work and introduced into the economy, and the effects of this are considered – lends itself quite easily to treatment in terms of a comparison between *before* and *after*.

This is the problem that both McCulloch (1821) and Wicksell, to take two examples far apart in time, really have in mind: the first with his theory of automatic compensation, and the second with his argument in favour of falling wages, put forward in a series of articles in the early 1920s and, in particular, in the one turned down by Keynes for publication in the *Economic Journal* in 1924 and recently published in the same journal (Jonung 1981). It is the same problem – with the attention already fixed on the point of arrival of the transition to a full use of the new technique – that Ricardo himself had in mind when, in his earlier view, he thought that 'no reduction of wages would take place' in consequence of the employment of machinery 'because the capitalist would have the power of demanding and employing the same quantity of labour as before, although he might be under the necessity of employing it in the production of a new or at any rate of a different commodity' (Ricardo, 1821, p. 387). The problem – as can easily be seen – is treated in terms of a shift of resources, 'the removal of capital and labour from one employment to another' (1821, p. 386), a shift that, when looked at from the point of view of a comparison between already defined situations, is *analytically instantaneous*.

In this perspective Ricardo's change of opinion – judged by Sraffa as the most revolutionary change in the third edition of the *Principles* (Sraffa, 1951, p. liii) – appears as a shift of focus from the effects of the use of machinery to the problems related to the phase of its introduction, rather than as a retraction of his previous opinion. His numerical example, which has an implicit sequential structure, helps in fact to throw light on what happens during the transition from a situation in which production is carried on without the help of machinery to a situation in which machinery is introduced. Unemployment appears as the result of a diminution in the 'gross produce' which can take place notwithstanding an increase in 'net income' (profits). A stage in which final output is less than it would have been if the change in the technique had not occurred is seen as the result of the conversion of circulating capital into fixed capital, implying a reduction of the Wages Fund, and hence of the employment that can be sustained, which will last until the greater productivity of the new technique builds this fund up again. The reason for a shrinking of the Wages Fund when circulating capital is converted into fixed capital is that the former comes back in the form of 'food and the other necessaries of life' (i.e., the items that make up the Wages Fund) after a single period, while this return takes longer when the circulating capital is invested in fixed capital. There will thus be a span of time, determined by a lengthening of the period required for the capital fund to return to its uninvested form, during which the Wages Fund will decrease and unemployment will appear.

Ricardo had a glimpse of this effect because in his example there is an implicit reference to the time dimension of the process of production, although it cannot be brought out explicitly because Ricardo himself, in drawing the implications of his example, falls back on the comparison between a situation in which part of the existing circulating capital has already been converted into fixed capital and the previous situation, thus abstracting from the process in time that has led from the one to the other. The hint at the discrepancy between investment at cost and investment of output capacity, thus, cannot be developed into a proper analysis of the phenomenon of technological unemployment and of the conditions required for it to take place,[1] since this analysis would make it necessary to consider explicitly the phase of construction of machinery, during which unemployment appears, while the comparison made essentially abstracts from it. The contrast between intuition and method is quite obvious, and focuses attention on the real analytical issue at stake when phenomena that imply in a relevant way a sequential occurrence of the events must be dealt with: the explicit consideration of a real time dimension within which a process of change can be properly articulated.

The 'Shifting Curves' Models

Ricardo's intuition made headway against the method of analysis he himself was using and depended essentially on the reference to the classical concept of capital as a fund, and on the implications of such a fund being invested in different forms characterized by different time dimensions.

Both the concept of capital as a fund and the hint at the relevance of the time dimension of production are absent from the neoclassical

---

[1] Before such an analysis could be adequately developed a century and a half was to pass. See Hicks (1970, 1973).

analytical framework, where the representation of the process of production and of technology, and the definition of capital as a stock of physical goods, are made perfectly coherent with an equilibrium analysis that fully relies on the method of the ·comparison. This framework, complete in all its details, lies behind the shifting production function model, which sees the analysis of technical progress as a comparison of different (successive) equilibrium positions, each representing a different technological state of the economy reflecting a different (superior) method of production. On the other hand, a method of production (a 'technique') is fully specified in terms of technical coefficients, which express particularly the proportion in which the existing resources must be combined in the process of production, thus giving shape to the capital goods (the 'machines') that embody the technique itself (see chapter 1, pp. 9–10). The techniques therefore have their own autonomous dimension, can be defined in a general and abstract way, and thus represented (as points of a production function), compared and classified. Technical progress is then identified with a modification of the technical coefficients that causes an upward shift of the production function.

The only problem for the economy, in this context, is to adjust its productive capacity to a new technique once this has been defined as *superior* according to some criterion (e.g. a higher rate of profit for a given wage rate) which generally implies the possibility of obtaining more of a given commodity (or of a given basket of commodities) from given resources. What matters – and is in the foreground of the analysis – is the point of arrival of the process of adjustment: the new configuration of the productive capacity that can be fully specified once the technological dimension of the technique as expressed by the technical coefficients is known. It is then all too easy to go straight to the conclusion, that is, to the comparison between *before* and *after*, abstracting from the (necessary and hence analytically irrelevant) process that will lead from one point, or path, to another.

What has just been said, however, does not depend strictly on the neoclassical analytical apparatus of production theory whose central concept is the production function, but extends to other approaches that, although differing from the neoclassical one in many aspects, share with the latter a representation of technical progress based on a similar notion of 'technique' and of the underlying process of production. Shifts in wage–profit curves derived from a Sraffian analytical framework (see, e.g., Schefold 1976, 1979) are expressions of the same approach, and allow the same kind of analysis, as shifting production functions: that is, an analysis made by means of comparison of points belonging to different curves.

In this context, to sum up, there is no room, and no need is felt, for a time dimension in which to order events vertically. Technical change, as has already been underlined in chapter 1, has the nature of a quantitative adjustment, and is brought about by a simple and analytically instantaneous shift of resources which makes it possible to define automatically the new productive capacity to be compared with the old one, characterized by a different combination of the same resources.

The Synchronic Representation of Production in Models with a Horizontal Sectoral Structure

The reference made above to the Sraffian framework helps to extend the argument developed in relation to shifting curves models more generally to models with a horizontal sectoral structure of production. True, multi-sector models give more information as to the working of the economy at any given point in time than do explicitly or implicitly aggregate models; however, they focus on flow balances – reflected in a transaction account for the whole economy conceived as being made up of separate industries – which imply already established and properly primed productive structures, and which, for that reason, allow only a comparison between different states of the economy under different conditions and/or at different points in time.

This is quite evident in input–output models 'à la Leontief', which are explicit flow models, but is also true of the Sraffa–von Neumann kind of models, which also involve stocks (of fixed capital goods) but transform them into flows by means of the joint production approach, according to which 'a process that uses capital equipment is regarded as a process that converts a bundle of inputs into a bundle of outputs; inputs are defined to include capital goods left over from the preceding period and outputs are defined to include . . . capital goods left over at the end of the current period' (Morishima, 1969, p. 89), so that the capital goods become a part of the gross output of the production process.

On the other hand the representation of the working of the economy given by multi-sector models relies on the implicit assumption that production is synchronized in such a way as to deprive the effective duration of the process of production of any analytical relevance. The flows considered in these models (the quantities produced and exchanged, the current prices at the time of the exchanges, . . .) are in fact transactions over a period of time that has a conventional length (the year, . . .) or is entirely arbitrary, as happens especially when the processes of production are standardized in such a way that each one is of unit time duration and 'those of longer duration may be considered as being composed of a number of standardized processes of unit duration, if we are prepared to enlarge our list of goods so as to include fictitious intermediate products' (Morishima, 1969, p. 91). The unit period can thus be made infinitesimally short, and the analysis can be carried out in terms of continuous functions of time.

Such an extreme disintegration of the process of production, however, is only possible when the effective duration of the elementary process no longer has analytical relevance – as is the case, for example, when the elementary processes are arranged in line in the factory system and enough of them are carried out that all stages of a given type of process take place at the same time (Georgescu-Roegen, 1965). In this case the vertical ordering of the successive phases of the process of production fades away, each stage feeding not only the immediately succeeding stage of the particular process to which it belongs, but also other stages belonging to other processes. The concept of the stage thus gives way to the concept of the sector, and it is the very nature of the concept of sector to be arbitrary and hence essentially divorced from any time dimension.

A synchronic representation of the process of production then becomes possible, implying that at each given moment a given product exists in all the stages of its fabrication. The instant the economy begins to work, the products start flowing out, and no interruption of its functioning can modify this state of affairs. Such an economy is capable of growing at a constant or even at an accelerating speed without waiting (Georgescu-Roegen, 1974). The production capacity increases simultaneously with the accumulation of producer goods and to the same amount: no trace is left, therefore, of the time articulation of the process of production or of the discrepancy between investment at cost and investment of output capacity on which Ricardo's example was based.[2]

Useful as they are for the analysis of an economy in a state of self-replacement or already geared to a steady growth, multi-sector models can portray a change, namely a technical change, only as a sequence of states of the economy 'depicting successive but separate levels of

---

[2] This synchronic representation is the logical counterpart of the underlying transaction accounting structure, according to which the performance of the economy during the given period is adequately represented by the flows observed within the period itself, whether these *ex post* values are interpreted as equilibrium values in the proper sense (i.e. as intersections of supply-and-demand functions) or, in a weaker sense, as only consistent with each other so as to ensure coherence between the building blocks of the system and hence economically feasible solutions (say, self-reproduction). On this point see Punzo (1984).

capital formed and labour trained', but cannot 'offer an insight into the intervening processes during which new capital is being formed and labour is being trained' (Lowe, 1976, p. 10). As Hicks pointed out with reference to the simplest case of sectoral disintegration, that is the case in which just two kinds of firm (those that make capital goods (the 'machines') and those that use them) are considered, 'the accounting division between Consumption and Investment is converted into an Industrial Division. There is investment while the machine is being built and there is disinvestment while it is being used. What therefore is liable to happen, if this method is adopted, is that the time taken to make the machine is liable to be forgotten' (Hicks, 1973, p. 5). However, the time taken to make the machine – machines being the concrete expression of the productive capacity embodying a given technique – is the relevant moment of the process of acquisition of innovation; and it is precisely the analysis of this process that, as we shall see, gets lost in models with a horizontal sectoral structure.

New Ideas in an Old Framework

The focus on already established structures of production and the essential abstracting from a time dimension along which to articulate a process of change on one side legitimizes the analysis of the change by means of comparison, and on the other reduces the change itself to a mere shift of resources which is in the nature of a quantitative adjustment.

This is immediately evident in explicitly or implicitly aggregate models – like one-commodity models or models where the production of a bundle of goods combined in fixed proportions is considered – or in multi-sector models that behave in an aggregate way, as is the case when a change in the relative weight of the different compartments of the economy results from an expansion (contraction) of supply and demand at the same pace within each sector but at a different pace between sectors. These models have in fact been clearly built to focus on the problem of the equilibrium between the demand generated by the economy and its supply capacity. A given productive capacity, the expression of a method of production in the sense of a given way of satisfying certain economic needs, can be operated at different levels of intensity (in the short-run) or augmented/reduced (in the longer run), in order to meet an increasing or decreasing demand. Whether the capacity is given or whether changes in its dimensions are considered, the problem is to produce *more* or *less* in order to fill a gap between supply and demand that appears at the level of final demand or in some intermediate stage of production. Technical change in this context – as we have already pointed out – is identified with the modification of given technical coefficients affecting the quantitative relation between given inputs and a given output, and thus implying reference to a productive structure that is and remains the expression of a given way of working of the economy.

The problem of a qualitative change that calls for a different solution to different productive problems can be seen more easily in multi-sector models where the structure of supply does not fit (or no longer fits) the structure of demand, and where this mismatch can be interpreted as a signal that what is required is not a simple change of the levels and/or the composition of the existing bundle of goods, but something else – which might mean some specific new goods or, more generally, something not yet well defined, as is usually the case when new needs appear and new ways of satisfying these needs are considered. What is required, in this case, is a different structure of production which, since it must reflect a different way of matching (perhaps different) economic needs, is not comparable with the existing one in terms of a *more* or a *less* that refer to an input/output relation that no longer has any meaning. From the comparison, which is in the nature of a quantitative adjustment, we must pass to the analysis of a qualitative change, that is to the analysis of the process through which the old structure of production is transformed into the new, different one.

The problem appears clearly in Pasinetti's model (1965, 1981), where changes in consumers' preferences at higher levels of real income result in rates of growth of demand that, in a multi-sector set-up, differ between commodities, so that the demand in some sectors cannot match the increases in productivity deriving from the learning by doing in production. The qualitative nature of the disequilibrium arising at the aggregate level is obvious. As the demand for some of the goods produced in the system gradually becomes saturated, the appearance of new products does not take place at such a pace that the aggregate demand can grow at the same rate as average productivity – because learning is slower in consumption than in production.

The first step to take, to focus on the process of qualitative change implied by such a set-up, is to cut the automatic one-to-one correspondence between the 'technique' and the actual structure of productive capacity, which makes it possible to deduce immediately the shape that the latter will acquire from the description of an innovation in terms of a modification of the technical coefficients. Pasinetti distinguishes in fact between the 'choice of technique function' listing the technologically feasible methods of production at a given moment of time, and the 'production function' referring to 'the actual production structure, which . . . may well be made up of various layers of different vintages (each layer representing the particular method actually chosen, at the corresponding time, from the "choice of technique function" of that time)' but represents 'the unique technical structure actually in operation' (Pasinetti, 1981, pp. 200–1). The step forward with respect to vintage models is represented by the fact that the actual productive capacity is no longer an array of sections each working (producing) in its own way, but is taken *as a whole* working in a *unique* way.

The production methods embodied in the existing productive capacity, and coming in succession through time from different 'choice-of-technique functions', are represented by vertically integrated labour coefficients, together with a series of complementary stocks of physical capital goods. These coefficients are the expression of the 'vertically integrated sectors' derived from Sraffa's subsystems (Pasinetti, 1973), which cut across the traditional industrial sectoral disintegration of the economy, going back to all the different productive links involved and permitting a description of the structure of productive capacity and of its functioning that can take account in a more realistic and complete way of the actual organization of the process of production. Changes in technology are introduced as movements of the production coefficients through time. These coefficients, however, are at each moment of time the result of an already realized process of adjustment that made the technological feasibility of the methods of production originally considered also economically feasible, in a given environment. The coefficients can in fact show· only *ex post*, that is when the productive capacity has already acquired its new, composite specification, and they have therefore no analytical role to play *during* the process of transformation of this capacity. If we look at them, and at their movements through time, we place ourselves once again in the *ex post* descriptive posture, which tells us nothing about the processes that brought about the existing technological alternatives and productive structures or about the specific forces that generated them.

The reason for Pasinetti being unable to draw the analytical implications of the qualitative aspects of the disequilibrium that arises in his model, and for his resort to the quantitative aspects of this disequilibrium – that is, the recurring divergences between supply and demand that make it impossible to maintain continued full employment – is that his vertically integrated model has essentially the same analytical structure as the standard input–output model, and so shows the same limits as the latter when it deals with the truly dynamic aspects of technical change. As the author himself takes care to point out, in fact, 'both models represent the same thing, looked at in a different way.

The difference, in other words, lies only in the classification, and we can pass from the one to the other by an algebraic rearrangement (a simple matter of computation) corresponding to a process of solving a system of linear equations; the production coefficients of a vertically integrated model turn out to be a linear combination of the production coefficients of the corresponding input–output model' (Pasinetti, 1981, p. 111).

THE ANALYSIS OF A PROCESS OF CHANGE

The Analysis of the 'Traverse'

The analysis of the 'traverse' – an attempt to take explicitly into account the transition from a given configuration of the productive capacity of the economy (equilibrium point or steady-growth path) to a different one – makes it easier to understand why this cannot be done properly in models with a horizontal sectoral structure (where, as we have seen, the transition itself comes down to an adjustment of a quantitative character to be realized through a simple reallocation of inputs between the different compartments of the economy), and how this depends mainly on the way in which the process of production is represented and the capital goods are dealt with.

Readjustment/transformation[3] of existing capacity is first and most easily analysed in models where all goods can be used indiscriminately as consumption or as capital goods, so a fall in consumption means automatically an increase in whatever capital goods are required to ensure the transition (Solow, 1967).

Heterogeneity between consumption goods and capital goods is introduced by Hicks (1965) in his two-sector (corn–tractor) model of the 'traverse', where tractors and labour produce both tractors and, used in different proportions, corn, and can move freely from one sector to the other. In this case changes in consumption no longer mean immediate opposite changes in the number of tractors, and the transition from one steady-growth rate to a higher (or a lower) one – the particular case examined by the author – can be realized only through transfers of labour and tractors between the corn sector and the tractor sector.

Technical change within a similar two-sector model is introduced in the form of a simple reduction of technical coefficients, implying

---

[3] The two terms can be used indiscriminately in most models of the 'traverse' where, as we shall see, they amount to the same thing.

that neither the consumption good nor the capital good changes its physical identity in the passage from one technique to the other; the 'traverse', as in the case of a change in the growth rate, boils down to a readjustment of the relative size of the two sectors to be realized through horizontal transfers (Spaventa, 1973).

Heterogeneity between consumption goods and capital goods, implying that when the quantity of a capital good is insufficient for the requirement of a new situation it cannot be increased by simply squeezing consumption, thus brings to light the problem of a process of transition. However, while the differentiation points to the process that the productive capacity must undergo in order to change its configuration, the representation of technical progress as a modification of technical coefficients and the hypothesis of transferability of the inputs reduces the change to a simple quantitative adjustment to be realized through horizontal transfers, thus cancelling the process that the analysis of the 'traverse' was intended to sort out.

In the attempt to focus on the process by which the productive capacity changes its physical form, that is, 'the formation, application and liquidation of that real capital which locks the economy into a particular technique', Lowe (1976, p. 10) moves a further step along the line of differentiation, which he assumes not only between consumption goods and capital goods, but also between the capital goods themselves. The economy is divided into three aggregate sectors, a consumer–good sector, say II, and two capital–good sectors: Ib, which produces the machines used as inputs in sector II, and Ia, which produces the machines used as inputs both in sector Ib and in sector Ia itself. There are therefore two different kinds of machines: those required to reproduce themselves and to produce the machines that will be used in sector II; and the latter ones, required to produce the consumption goods. Only the machines of the first kind can be transferred, and they can be transferred between the two capital–good sectors but not between the capital and the consumer–good sector, as in traditional two-sector models.

Implications of this set-up go far beyond the simple addition of a further sector to the model. Differentiation between the capital goods, and the limit set to their transferability, in fact introduces a new dimension into the process of production, establishing a sequence (Ia–Ib–II) according to which fixed capital in sector Ia must be increased *before* any increase can be obtained in the production of consumption goods, so the transition can no longer be accomplished through mere horizontal transfers between sectors. With the sectors beginning to lose their character of simple compartments of the economy and tending to become different phases of a process of production articulated sequentially – that is, an operation opposite to the one carried out in standard sectoral models, mentioned earlier in this chapter (pp. 21–2)—there emerges the problem of a *vertical* transfer of resources between different phases, to be carried out in a context in which the timing of the events matters.

However, while on one side the model implies a sequence, it also retains circularity within the capital–good sectors, where the equipment considered is technically suited to reproduce itself (in sector Ia) as well as to produce (in sector Ib) the equipment required as input in the consumer–good sector; and circularity causes the effect of the differentiation between the capital goods to vanish. In fact, according to Lowe, it is an expansion of sector Ia at the expense of sector Ib that starts the transition, and thus we are back at the orthodox sectoral representation where the adjustment to a new situation is realized through horizontal transfers between sectors.

Circular Relations in Production and the Problem of 'the Time Taken to Make the Machine'

The fact that the 'traverse' is accomplished through a reallocation of capital goods between the different sectors entails, we have pointed out, that when an innovation is considered the old and the new technique continue to use the same inputs, although in different proportions, and that technical progress, in effect, does not require a qualitative transformation of the productive capacity but a simple reshuffling, which analytically is an instantaneous operation. Technical progress, therefore, is adequately analysed by means of a comparison between equilibria – each exhibiting a productive structure already adjusted to the dominant technology – while the process of transition in the sense of 'what happens on the way' is lost to sight.

However, if we reckon that the relevant aspect of the acquisition of new technology is the very process through which the productive capacity is transformed so as to acquire a new specification, and that this transformation requires the liquidation of old equipment and the building of new and different capital goods,[4] then this problem cannot be dealt with properly in models with a horizontal sectoral structure where, we repeat, 'the time taken to make the machine' is liable to be forgotten. In fact, when such a transformation is considered the

---

[4] It is worth underlining the fact that the building of new machines 'stylizes' a reshaping of the productive capacity which must in effect be looked at in a more general way as involving, besides (or even in the place of) the machines, a whole series of other elements (human, organizational etc.).

problem is not so much to liberate the capacity to produce *more* as to build *different* productive equipment, that is – getting back to Lowe's model – not so much to *expand* sector Ia at the expense of sector Ib, as to *change the way it works*. What does the trick in multi-sector models, that is what reduces the *different* to the *more*, is the assumption that *old* machines can be used to produce the *new* ones. Thus circularity can be retained also in the presence of technical progress and the reshaping of productive capacity realized once more through simple transfers between sectors.

Old ways of producing new equipment, however, clearly mean that no change in the process of production takes place in the relevant moment of the embodiment of the new technology, that is in the phase of the making of the machine – and this is why a simple expansion of one sector (namely sector Ia) can be smuggled in as that change in the process of production that technical progress implies and that on the contrary is left outside the door. The assumption that old equipment, and old ways, can be used to produce the new equipment disposes of the process through which productive capacity acquires a different specification in the easiest way: by cancelling it (Amendola, 1984a).

It is not enough, then, to state that the phase of the embodiment of a new technology must be taken explicitly into consideration; we must also rely on an analytical framework that allows us to deal with the process of embodiment, that is what makes it possible to articulate it (as all processes should be) in real, irreversible time. This is the lesson that we draw from the analysis of the 'traverse', and in particular from Lowe's model, together with the perception of the strict relation between the sectoral representation of the economy (with its synchronic representation of the process of production) and the hypothesis of transferability, and of the loss of relevance of such a representation as the capital goods become more and more specialized and hence less and less transferable.

The Neo-Austrian Approach

Circular relations in production, although useful for an understanding of the working of an economy with a given structure of production, are thus an obstacle to the analysis of the process of innovation when this implies a structural change. In fact a horizontal sectoral structure of the model carries with it the view of capital goods that can take part in different types of processes of production and hence exist in their own right, with their own physical specification. This, as we have seen, reduces the process of change to the simple and analytically instantaneous shifting of existing equipment to a different task and/or to a different compartment of the economy.

So as to shed light on and to be able to analyse the.phase during which the transformation of productive capacity takes place, we have to free it from the encumbering presence of capital goods that can serve new as well as old purposes. Capital goods, in other words, must be dealt with in a different way; this, in turn, requires a different way of conceiving the process of production, with a return, in particular, to the idea that 'production takes time'.[5]

In his Neo-Austrian model, which makes it possible to exhibit the intertemporal complementarity and the unilateral dependence of the successive events on the preceding ones that characterize the phenomenon of production, Hicks (1970, 1973) goes for total differentiation and non-transferability. Labour is assumed to be the only freely transferable original input of a process of production defined as a scheme for converting a stream of labour inputs into a stream of homogeneous final output (consumption goods). The process is fully integrated vertically and must be taken as a whole over time. The capital goods become the particular expression of each kind of process, within which they are produced, cannot exist outside it and hence cannot be transferred: they are implied, but are regarded as intermediate products and not explicitly shown. The sectors disappear completely; real, irreversible time emerges as the fundamental dimension of the process of production and the time articulation of the different phases (the 'time profile') becomes its distinctive feature.

In this context capital is no longer identified with a stock of physical capital goods; the latter are seen in fact as a result (although an intermediate result) of the process of production, while capital – going back to the classical tradition kept alive by the Austrian theory of capital (Böhm-Bawerk, 1889; Wicksell, 1901; Hayek, 1941) and essentially shared by Schumpeter (1934) – is conceived as a fund, namely a Wages Fund that in the model is made up of the final output available for financing the labour required for starting the processes of production and for carrying them out. The attention can thus be shifted from 'the machine' to 'the making of' the machine. Only labour, and the

---

[5] It will be remembered that the time dimension of production and the concept of capital as a fund were the two relevant analytical elements of Ricardo's example on the problem of the appearance of technological unemployment. Both of these elements have disappeared, together with the problem that they helped to clarify, from the theoretical models that have subsequently dominated the scene. It should therefore be no surprise that the attention goes naturally back to these same elements when one tries to bring back to light that process of change with which the phenomenon of technological unemployment is essentially associated.

resources required to employ it,[6] will in fact be present on the stage at the moment of the change from one method of production to another; it will thus be possible to consider a completely new start to the process of production along different lines, and thus to take explicitly into account and to analyse the phase during which the process itself will acquire its new profile and will build *its own* specific equipment in its own *original way*.

The analysis is carried out with reference to a process of production that extends over a sequence of periods (weeks) that integrate two successive phases: the phase of construction, during which the machine embodying the new technique is built, and following that, the phase of utilization of this machine. 'The one week relations . . . determine the course of the model in week T, when everything that has happened before week T is taken as given. Having determined the course in week T, we can then proceed to week $T + 1$, applying similar relations, but with the performance of week T now forming part of the past. And so on, and so on. The path of the economy, over any number of successive weeks, can thus be determined' (Hicks, 1973, p. 63). In particular, the output and the employment, at week T, depend entirely on the processes that have been started in the past, and on the techniques that are used in those processes. The basic element of the sequence is the rate of starts of new processes, which is endogenized and made dependent on the rate of starts in the past. This is obtained by assuming on one side that all output that is not consumed is invested (the hypothesis of 'Full Performance' according to which activity is limited only by savings) and on the other side that consumption out of profits, the take-out that cannot be used to start new processes, is constant in absolute terms. Once the rate of starts is made endogenous, the model becomes sequential and can be used to describe the 'traverse' from one steady-growth path to another, characterized by the use of machines embodying a *superior* technique.

In this model the relation between the phase of construction and the phase of utilization of the new capacity, which determines the time profile of the process of production, acquires a great importance. It is in fact what happens during the phase of construction, and the time required by the latter before the utilization of the new machines can begin, that makes it possible to sort out the relevant phenomena of the process of transition between techniques. In particular, the highly controversial issue of technological unemployment can be dealt with in a systematic way, and a proof of Ricardo's 'machinery effect' (the

---

[6] Physical resources (i.e. final output just referred to) or financial resources, as in the model proposed in the next chapters.

introduction of machinery leads to a temporary contraction in final output and has an adverse effect on employment in the short run) is provided for the particular case of an increase in mechanization of the processes of production (the use of more fixed capital in order to economize on circulating capital) implying the use of more labour near the start of the process in order to use less later.[7]

From the Embodiment of the Technique to the Creation of the Technology

The Neo-Austrian model highlights the time structure and the technical intertemporal complementarity of the process of production, thus providing the framework for the analysis of the making of the machine as the relevant moment of the process of innovation. The way capital goods are dealt with, on the other hand, prohibits the reduction of this process to a piece of telescoping and opens the way for it to be treated as a sequential process.

The analysis, however, is developed by Hicks with reference to a barter economy with a single homogeneous final output, where the hypothesis of Full Performance implies a flow equilibrium in each period, in the sense that the productive capacity represented by the stock of processes inherited from the past is always fully matched by the existing demand.[8] The adjusting variable is the rate of starts: all the output not absorbed by consumption out of wages paid to labour engaged on old processes (whether still in the construction or already in the utilization phase) and not absorbed in consumption of other kinds, is in fact used for starting new processes. However, current output is largely predetermined by decisions that have been made in the past, so the rate at which new processes are started depends ultimately on the rate at which they were started in the past. Thus, given the parameters of the model, the 'traverse' becomes a predetermined sequence of decisions of a quantitative kind which can be fully traced out *ex ante* and where expectations play no role; along it, in fact, there is neither real choice as to *how much* to invest nor as to *how* to invest the preassigned amounts, since the very definition of a *superior* technique automatically implies its choice.

---

[7] For an extension of the 'machinery effect' to all cases of technical progress in processes of production with a more general time profile, in a Neo-Austrian model, see Amendola (1972).

[8] The accumulation of stocks during temporary failures from which there is expectation of sudden recovery, or the carrying of *normal* stocks to avoid the extra costs of small orders or to meet a minimum degree of uncertainty, can be introduced without substantially altering the argument (Hicks, 1973, pp. 52–3).

This, however, should come as no surprise. Although concentrating on 'what happens on the way' – that is, on the making of a fully specific and non-transferable machine that has its place *inside* the process of production, and on the implications thereof – the Hicksian 'traverse' in fact still portrays the process of innovation as the embodiment in the productive capacity of the economy of a *given* technique, to be realized through a gradual diversion of resources (from the production of the old to the production of the new machines) that, although spread out in time, retains the character of a quantitative adjustment. The difference, as regards the model with a horizontal sectoral structure, is that the reallocation concerns generic labour – in order to set up a process of embodiment of a new technique that can be exhibited with all its characteristics and implications – and not particular capital goods, the simple shifting of which to a different task and/or compartment of the economy would already imply (the embodiment of) the new technique.

True, this makes an important difference, and is a great step towards the analysis of innovation as a process; but something more is required if we wish to interpret this process as a qualitative change, that is as the creation of new technology that brings about different productive structures rather than the absorption of given technological advances into the prevailing productive structure.

We focused in chapter 1 on the concept of technology as an environment characterized by skills and qualifications that make it possible to devise and implement different solutions to different productive problems, and we have pointed to a heterogeneous human resource as the expression of this environment. The process of innovation has then been defined as a process of research and learning that brings about new and different skills and qualifications and thus implies a modification of the characteristics and of the structure of this resource. What we need, then, in the first place, in order to deal with this kind of process, is the substitution of a heterogeneous and specific human resource for the homogeneous and generic labour input contemplated in the original Neo-Austrian model. Formally, this can easily be done by substituting a vector for a scalar in defining the existing resources, and a matrix for a vector in defining the different inputs required in each period of the process of production (details are given later in chapter 4). What is implied, however, is not only this: it is a shift of attention from a 'labour input' considered for its employment in a given process of production to a 'human resource' considered for its capacity to think and implement new and different processes – in correspondence with the shift of attention from a given process as the embodiment of a given technique to technology (and its changes) as a potential source of different kinds of processes.

We assume therefore that the heterogeneous human resource that is at each moment the expression of the technology of the economy is *specific*, in that it embodies characteristics (and potentialities) that are the outcome of a process of learning that can take place only as the result of an innovative choice: that is, of the choice to set out along an innovative path, starting processes of production whose profiles are not yet well defined and that will acquire a precise contour while being carried out, due to the growing acquaintance with the productive problems involved.

To sum up: the Neo-Austrian model assumes that capital goods are fully specific and that their specialization (the embodiment of the 'technique') takes place inside the particular process of production corresponding to each given technique. We take a further step assuming that the human resource also undergoes specialization while the process of production is being carried out (provided it is an innovative process where learning is taking place), and that this specialization consists in enlarging its potentialities rather than in adaptation to a specific task. The creation of technology, identified with the specialization and the enrichment of the human resource, is thus brought inside a process that, thanks to the model, can be taken explicitly into account and analysed.

On the other hand the process of innovation, in the perspective of a Neo-Austrian framework so modified, no longer has the character of a predetermined succession of decisions of a quantitative kind leading to a given point of arrival, but must be looked at as a related sequence of *true choices* that have to be reconsidered in the uncertain context of an environment that is expected to change (and from which, therefore, more information is expected to come) as the result of each successive step. The decisions taken at each given moment, in fact, reflect (and are constrained by) an environment qualified essentially by a given structure of the human resource, but contribute at the same time to the modification of this environment by feeding a process of learning that will result in an enrichment of the human resource as a potential source of productive developments, and thus in the opening of new paths along which the subsequent steps can be taken.

We are therefore faced with the intertemporal complementarity of a decision-generating process which must be incorporated into our Neo-Austrian model to outline – together with the technical intertemporal complementarity of the process of production – the profile of a process of innovation interpreted as a process of creation of technology. Such a process, on the other hand, is characterized mainly by sequential learning; it calls therefore for a truly sequential analytical framework, in which, as we shall see, money plays an essential role.

3

The Sequential Analytical Framework

SEQUENCE ANALYSIS

Sequence Models

Sequence models 'are constructed out of certain relations in time assumed to be fundamental' (Lundberg, 1937, p. 51).

First of all, regard is paid to the fact that production takes time, and this makes it possible to highlight 'the connections in time between income and spending, between input and output, between a change in receipts and the corresponding change in production activity' (1937, p. 48). In particular, consideration of the time dimension of production implies that the costs paid out for the production of consumers' goods are not necessarily used for the purchase of the same goods. However, this discrepancy between the costs incurred for producing the final output of a certain period and the income used to purchase the same output cannot arise in a barter economy (or in a virtual barter economy, in which money is only a unit of account that represents the product) where the only thing acceptable in exchange for today's product can only be more of today's product.

It is then not by chance that sequence analysis was developed with reference to monetary economics by Swedish economists, starting from Wicksell. As is well known, Wicksell considered a monetary stationary equilibrium where the equality between the market rate of interest (the cost of borrowing money) and the natural rate of interest (the profitability of borrowing money for investment) emerges as a condition for money price stability. In such an equilibrium we have a stationary process, with a constant flow of money matching a constant flow of goods, determined by fixed technical conditions; the time lag represented by the production period is then analytically irrelevant, and production costs and purchasing power can be considered as simultaneous.

They are *not* the same thing, however, as can be easily seen when the economy suffers a shock that causes the potential discrepancy to appear, and the economy to react in a sequence of successive changes. This is what happens in particular in the cumulative process of change in the price level, considered by Wicksell in *Interest and Prices* (1898), as a result of an increase of the real, or natural rate of interest above the nominal, or market rate. The consideration of the production lag implies in fact that the costs incurred for increasing production in order to reap the higher profits brought about by the difference between the real and the nominal rate of interest are not covered by a corresponding output, since the 'operations of the producers cannot be supposed to become adjusted immediately to their own effects' (Lundberg, 1937, p. 56). We shall thus have an initial increase in the prices of the production inputs, followed by an increase in the prices of the consumer goods, which will re-establish the difference between the real and the nominal rate of interest, thus feeding a cumulative process of price changes.

Wicksell's analysis deals at the outset exclusively with the current production of consumption goods, and his habit of working with a circulating capital model implies considering self-contained periods that are characterized by different price levels but are not linked together in a true sequence (Hicks, 1965, pp. 60–1). Later on, when the formation of durable capital goods is considered in the *Lectures on Political Economy* (1901), the analysis 'loses precision with regard to the timing of the events' (Lundberg, 1937, p. 57). Notwithstanding the difficulties encountered and the shortcomings of the analysis, however, 'there remains *the technique* for exploring an expansion (or a contraction) process . . . a technique which can be extended by taking more and more factors into consideration' (1937, p. 58).

Moving along this line, that is, with the original aim of explaining changes in the price level, Lindahl (1930) manages in fact to develop a proper sequential analysis. With his temporary equilibrium method he reduces the process of change to a sequence of periods, such that within each one of them change can be neglected, and he explicitly introduces expectations as independent variables in the determination of the single-period equilibrium. A transaction accounting structure provides the framework for the analysis of the single-period (Hicks, 1956), which is carried out in terms of a comparison between plans (unchanged over the period) and realizations, and in which the equilibrium between supply and demand, given the expectations, is ensured by flexible prices which take care of the possible gaps between plans and realizations.[1]

The changes (and, in particular, the alterations of the plans) take place at the junction of successive periods, which must therefore be strung together in a sequence.[2] It is again the comparison of what does in fact happen with what is expected to happen, by focusing on the role of expectations, which provides the link, namely assuming that the plans that will determine the events of a given period reflect expectations based on the results of the preceding period(s).

The connections, over successive periods of time, of plans and decisions – reflecting essentially the ways in which expectations are formed – therefore represent the causal element of the sequence. If the plans are fulfilled it is possible to postulate the existence of an expectation function of constant form, which indicates that the method of arriving at the expected values of the relevant variables remains unchanged. We then have an equilibrium sequence analysis. This does not necessarily imply, however, that the actual values always strictly coincide with the expected values; small divergences can in fact be considered, provided the situation is sustainable (Lindahl, 1934), or, in a somewhat different but similar interpretation, provided 'the actions of the agents . . . do not prove to be systematically and persistently inconsistent' (Hahn, 1974, p. 61).[3]

The Breaking of a Sequence

Essentially, the above framework can be retained to highlight the determinants of an innovative choice interpreted as a qualitative change

---

[1] In contrast, if fixed prices are considered (as Lindahl himself did later (1934, 1939 part I) under the influence of Myrdal (1933)), demands and supplies within each period are not necessarily equal, and this can result in an unwanted accumulation or decumulation of stocks. Changes in prices then take place at the transition between consecutive periods, as a consequence of what has happened to stocks and orders in the previous period(s).

[2] The analysis of the way in which successive periods are linked in a sequence has been called a 'continuation theory' which, together with the 'single-period theory', defines a dynamic analysis interpreted as analysis of processes of economic change (Hicks, 1956).

[3] On the other hand a disequilibrium sequence analysis is considered by Lundberg (1937), who assumes given response functions, that is, relations between successive periods that reflect routine behaviour and that keep the same form independently of whether the expectations are fulfilled or not. Only short-run models for discrete periods of time can be considered in this case, however, since it is realistic to assume that a 'routine is tried for a certain period of time, but, if found inappropriate, is replaced by another one' (Hahn, 1952, p. 27), and that a new model will then have to be set up.

and to analyse the decision-generating process which is the essence of the ensuing process of innovation as we have defined it.

The first point to stress, on this subject, is that a qualitative change conceived as the search for a new way of matching (perhaps new) economic needs, is always to be seen as the breaking of a sequence. An equilibrium (or a disequilibrium) sequence characterized by an expectation function of a constant form (or by a given response function) implies in fact that there is no change in the model/theory of the working of the economy referred to, i.e. that we are in equilibrium according to Hahn's definition: 'an economy is in equilibrium when it generates messages which do not cause agents to change the theories which they hold or the policies which they pursue' (Hahn, 1974, p. 25). This is certainly no longer true when the case for an innovative choice arises.

For the breaking of a sequence to be put properly into focus within the Neo-Austrian framework referred to, we need in the first place to consider a distinction between short-term and long-term expectations.

Short-term expectations extend over a single period – which can be made to coincide with the time required for a round of final production to be carried out – and concern the amount and the composition of the (demand for the) final output to be obtained from the existing productive capacity. They can be assumed to depend on past experience in the way already mentioned.

Long-term expectations, on the other hand, refer to decisions that do not concern the degree of utilization of the existing productive capacity, but a modification of it, implying the starting of processes of production that will reach the phase of utilization in future periods, after the phase of construction. These expectations depend on the degree of confidence of the agents (producers and consumers) in the information they have, that is, on the variation they are anticipating in their beliefs. When the agents have fully adjusted to a given way of working of the economy as synthesized by the structural parameters referring to the strictly related aspects of the technology and of the consumers' preference system, the long-term expectations will reflect the degree of confidence in the existing state of affairs which will result in a given structuring of the patterns of production and consumption (steady growth, of course, being the example of an equilibrium sequence that comes immediately to mind). In such a situation the short-term expectations will also be consistent with the long-term expectations in the sense that, although at a different level, they will share the confidence reflected in a given way of getting and processing the information from the environment.[4]

---

[4] Non-fulfilment for one or more periods will not necessarily imply a change in the way these expectations are formed, as was pointed out in the last section.

When there is the feeling or the perception that something new and different,[5] although not yet clearly specified, is going or needs to happen, uncertainty (or a greater degree of uncertainty) appears, associated with the prospect of learning. The more the agents expect to learn, the greater is the decrease in the confidence they have in the existing state of affairs, because the greater is the likelihood of a substantial revision of this in the future. There will therefore be an increase in the variation they anticipate in their beliefs (Jones and Ostroy, 1984), which can cause a break in the sequence, if it means a loss of confidence in the existing 'model' of the economy. A modification of the long-term expectations, no longer formed in the same way, will then follow, reflecting the appearance of a case for qualitative change:[6] it will in fact be a signal that the agents are considering the advisability of substituting a new model for the old one and are hence placing themselves in a position of learning, that is, in a position in which the new model is not yet in place and the new relations on which to rely have still to be learned.

This will result in a search for flexibility, expressed in the first instance by an increase in the demand for flexible positions – that is, positions allowing for waiting, postponement or quick revision of decisions – which will be the most immediate consequence of the uncertainty associated with the breaking of the sequence. However, this demand for flexibility 'is basically unconnected with risk aversion . . . flexible positions are attractive not because they are safe stores of value but because they are good stores of options' (Jones and Ostroy, 1984, p. 14). When flexibility, in this sense, is associated with 'the cost, or possibility, of moving from an initial position to various second period positions' (1984, p. 16), holding liquid assets appears as the most appropriate answer. Liquidity, in fact, 'is freedom. When a firm takes action that diminishes its liquidity it diminishes its freedom; for it exposes itself to the risk that it will have diminished, or retarded, its ability to respond to future opportunities' (Hicks, 1979, p. 94). Any long-term commitment like an investment in fixed capital whose 'illiquidity' is measured by the time required to complete a technologically irreversible process certainly constitutes such an action.[7]

---

[5] Different needs, or different ways of satisfying them, implying a restructuring of production and/or consumption according to different patterns.

[6] The important thing to understand, in economic theory, is how expectations change and not what expectations are.

[7] This is all the more true in a Neo-Austrian model, where the hypothesis of full vertical integration implies that every process of production must always be taken as a whole over time – that is, once started, must always be brought to its end.

The Search for Flexibility, Liquidity and the Role of Money

Accrued preference for liquidity, as the attribute of assets whose acquisition can be easily revoked as opposed to assets that on the contrary imply commitment to a given course of action, emerges therefore in the first instance as the most relevant aspect of the search for flexibility due to the breaking of a sequence.[8] However, such a position cannot be translated into an adequate form of behaviour unless we can rely upon financial reserve assets, the most important of which is certainly money.

We have already mentioned the role of money – when it is considered in its own right and not simply as representing product – in making it possible to bring out the discrepancy between production costs and purchasing power as a first implication of the consideration that production takes time.

Money, however, plays an even more important role in sequence analysis when learning is considered.[9] As was pointed out by Hahn (1973), in fact, money becomes essential in real sequence economies, that is, in economies where there is a sequential learning that implies a decision process conceived as a related sequence of choices and not as a predetermined succession of decisions that, while referring to transactions at different dates, are essentially independent of these dates (Radner, 1968) and hence are not sequentially dependent on each other.[10] When the sequence of periods makes it possible to get more information on the environment than is available in the first period, money becomes relevant less as a contingent store of value than for the character of liquidity attached to it which, as we have seen, emerges when the future alternatives associated with a given choice are the most important aspect of the latter.[11]

Both the problem of an innovative choice, considered as the breaking of a sequence, and the analysis of the ensuing process of innovation

---

[8] This is true not only of the producers but also of the consumers. In fact 'households do not place (future) contracts because they prefer to keep their options open in the uncertain world they live in. . . . In general consumers can avoid future contracts because they can choose a generalized form of command over future goods: money' (Brunner and Meltzer, 1971).

[9] As is the case in the analysis of a process of innovation as we have defined it.

[10] Dating brings in the future, but does not necessarily bring in the effects of the passage of time. The temporal order of decisions, in fact, matters when information is incomplete and is expected to change, and people learn from realizations that they could not have anticipated.

[11] 'For liquidity is not a property of a single choice; it is a matter of a sequence of choices, a related sequence' (Hicks, 1974a, p. 38).

interpreted as a sequential process of learning, call therefore for the presence of money, which, for simplicity, we can assume to be the only financial asset.

Dealing with a monetary and not a barter economy, on the other hand, adds a sequence of exchanges, which takes place within each period, to the sequence of production, which extends over successive periods. The wages are paid, and the exchanges are made, in fact, in money terms, and hence the resources required to carry on the processes of production are financial resources, not physical output. At the beginning of each period the resources available for advancing the wages to the workers employed in the processes of production (the Wages Fund)[12] may have an internal and/or an external source. The internal resources come (a) from the proceeds of the sales in the previous period(s), and (b) from a reduction (if any) of the proportion of money held by the producers as a liquid reserve asset, and a corresponding increase in the amount of it employed in the processes of production.[13] The external resources are represented by the money obtained by borrowing outside the economy.

The current output obtained with the existing productive capacity will be matched, at the end of the period, by a final demand coming from (a part or all of) the wages advanced to the workers who have taken part in the processes of production, plus the money kept by the producers for their own consumption, plus (or minus) the funds coming from a reduction (increase) in the proportion of money held by the consumers as a liquid reserve asset.[14]

Capital, in this context, is a fund made up of money, but it does not coincide with the latter. It is in fact represented by that fraction of the financial resources (money, in our case) that is actually made available for carrying on the processes of production.[15] The changes in the proportion of money devoted to production and/or consumption thus acquire a great analytical relevance: they are in fact not only an

[12] And for financing the consumption of the producers.

[13] That is, in Keynes's terminology, from a change in the distribution of the quantity of money owned by the producers between Financial circulation and Industrial circulation (Keynes, 1930, chapter 15). It goes without saying that a change *in favour* of the Financial circulation will *reduce* the resources available for financing the processes of production and consumption out of profits.

[14] What has just been said about the changes in the distribution between the Financial and the Industrial circulation holds for the consumers in the same way as for the producers.

[15] This recalls Schumpeter's definition 'Capital is nothing but . . . a means of diverting the factors of production to new uses, or of dictating a new direction to production' (Schumpeter, 1934, p. 16).

important aspect of the behaviour of producers and/or consumers immediately following the breaking of a sequence, as we have just seen, but also a main determinant of the effective evolution of a process of innovation, as will be shown in particular in chapters 4 and 5.

Flexibility and the Nature of the Learning Process

An increase in the demand for liquid assets, as we have seen, is the most immediate expression of the search for flexibility following the breaking of a sequence and the appearance of uncertainty (or greater uncertainty) about the environment associated with the prospect of learning.

However, when the nature and the sources of the process of learning involved are considered, the problem appears in a different light.

Holding liquid assets, with the possibility of the postponement of detailed and binding decisions that this implies, may in fact be a correct response when learning concerns existing opportunities, about which more information is expected to appear, or the sequential arrival of attractive new opportunities as the result of the mere passage of time. These opportunities, in other words, do not depend on the decision makers' actions and can be expected to become available independently (or in the absence of) such actions; furthermore, it is assumed that, say, the improvements in the technology brought about by someone else's actions can be appropriated freely (or at a given cost).[16]

Flexibility, in this case, has a defensive (static) significance: '*not to diminish* the options for the future'. A flexible choice is then a choice that does not reduce the future (given, or exogenously accruing) alternatives associated with it; the most liquid choice, in this perspective, is also the most flexible one.[17]

When the focus is shifted from the adoption/diffusion of exogenously accruing technological advances to the source of these advances – identified, in our interpretation of the process of innovation, with a process of learning that brings about an enrichment of the human

---

[16] This is the hypothesis behind Rosenberg's argument on 'technological expectations' concerning the timing and the significance of future improvements, according to which 'a firm may be unwilling to introduce the new technology if it seems highly probable that further technological improvements will shortly be forthcoming' (Rosenberg, 1982, p. 107). It is worth noting that this argument holds when both the *new technology* and the *further improvements* are considered as given for the firm, and the attention is on the *adoption* and the *diffusion* of exogenously determined technological advances.

[17] Jones and Ostroy (1984, p. 17) define as perfectly flexible those positions from which 'all second period positions can be reached with no switching costs'.

resource and of its capacity to devise and implement new solutions to new productive problems – and when this process is seen as the result of research and experimentation that can only be carried out by actually taking an innovative path, the whole perspective changes. The acquisition of entirely new and different skills and capacities – unlike the improvements represented, for example, by the design of a machine, which anyone can appropriate, although at a cost – appears in fact as a kind of learning which can only be the result of doing, not of waiting.

Holding liquid assets, in this case, is no longer the appropriate response to a search for flexibility because the meaning itself of flexibility changes together with the nature of the learning process considered. Flexibility, in other words, acquires an active (dynamic) character: 'to *increase* the options for the future'; and a flexible choice will then be not so much a choice which does not diminish the capacity to respond to oncoming opportunities as a choice that will *itself* bring about new opportunities, thus enlarging the gamut of future options.

The counterpart of this shift from a passive to an active interpretation of the concept of flexibility is the passage from the passive posture of waiting, which finds concrete expression in an accrued preference for liquidity, to the active posture represented by an innovative choice.

We can therefore interpret the increase in the demand for liquid positions that immediately follows the breaking of a sequence as the behaviour of an agent who, while having doubts on the existing model (i.e. on the way the economy is working),[18] is not yet fully convinced that the model must be discarded and replaced by something new and completely different. When (and if) such a conviction is reached,[19] the adequate response to the search for flexibility – in the light of the nature of the phenomenon of learning associated with our interpretation of the process of innovation – can be but an innovative choice, and the most important problem then becomes that of the viability of such a choice, strictly related to the determinants of the evolution of the process of innovation.

THE ANALYSIS OF A QUALITATIVE CHANGE

Stock Disequilibrium

We have seen how money introduces a sequence of exchanges within each single period. It is now time to consider the effects of the

---

[18] Doubts such as to prevent the agent from taking part in it in the way he or she had done before.
[19] Which is not always the case, and might take some time even when it is.

introduction of money on the sequence, that is, on the (nature of the) links that come to be established between successive periods. We shall then be able to put the process of innovation into its proper light – and we shall see that it is the very presence of money that makes it possible to conceive such a process as a qualitative change, and that it is this presence that, at the same time, determines the nature and the actual articulation of the sequence through which the process itself takes place.

The case for an innovative choice interpreted as a qualitative change clearly arises when the stock of the processes of production actually carried on – that is, the productive capacity of the economy shaped according to a given technology which has its expression in the strictly related structures and patterns of evolution of production and consumption – becomes a disequilibrium stock, that is, a stock no longer adjusted to the existing expectations.[20] This, on the other hand, must be taken not so much in the sense that the existing productive capacity is no longer adjusted to current demand (or to its expected growth) from a quantitative point of view,[21] as in the sense that it no longer represents the *right* answer to changing, although not yet clearly specified, requirements; that is, from a qualitative point of view.

The appearance of a stock disequilibrium of this kind, as a result of a modification of long-term expectations no longer formed in the same way, will on the other hand automatically bring about a flow disequilibrium (an inequality between current demand and current supply) since the plans for the current period, based on short-term expectations that are no longer consistent with the long-term ones, will not be realized.[22] The stock disequilibrium will thus be interpreted as the result of a flow disequilibrium due to a mistake that can be corrected, and not as the signal of a change in the way

[20] We have already pointed out that such a situation does not necessarily have to originate from the demand side, as a result of new, although not yet clearly expressed, needs of the consumers, but can also be the outcome of a feeling on the part of the producers that new methods of production are about to appear, or should be sought out, that entail a restructuring of production and consumption.

[21] In which case the problem would be just to increase or to decrease the capacity itself (or some sections of it) in order to make it match a greater or a smaller demand (or a different composition of the demand).

[22] Thus, to remain within the framework proposed, a modification of the consumers' preference system resulting in an accrued preference for liquidity will render the existing stock of processes of production a disequilibrium stock in the above sense, and will at the same time bring about an excess supply of final output in the current period.

of forming the expectations reflecting structural modifications that call for new answers to (new) problems, that is, for a different model.[23]

Whether the flow disequilibrium is (rightly) interpreted as the consequence of an underlying stock disequilibrium or (wrongly) as its cause, however, it should not be thrown over onto the price side. We want disequilibrium stocks (real and/or monetary) to appear and to be present; they are essential, from the analytical point of view, in order to outline a sequential process of change. 'Any economic entity which is left in a state of disequilibrium will take steps to right that disequilibrium; that is the characteristic effect of disequilibrium; it is the way in which disequilibrium carries its effects down the sequence' (Hicks, 1965, p. 82). That is why, in what follows and particularly in the model described in chapter 4, we shall make a 'fix-price' assumption – that is, we shall assume that prices, all prices, are fixed at the beginning of each period and kept unchanged throughout it, whatever the market conditions, and that they can change only at the junction from one period to the next.[24]

Carrying the Disequilibrium down the Sequence in a Monetary Economy: the Financial Constraint

Disequilibrium stocks are then essential for sketching a sequential process of change. We want something more, however; we want the change to be interpreted as a qualitative change, and to take place *inside* the process of innovation so as to be able to deal analytically with it. When the qualitative change takes place *outside* the process, in fact – as is the case when the appearance of a new technology (whether fully or partially developed) is assumed to be the precondition of the process itself – this comes down to a mere quantitative adjustment (see chapter 1).

Now, here it is where money comes in. It is the presence of money, in fact (that is, the possibility of considering money stocks), that permits

---

[23] An excess supply, as just mentioned, will leave the producers with a stock of goods that they do not wish to hold. If these goods are perishable, and hence go to waste, the funds that were invested in them will be lost, and the producers will be left with a debt that they are not willing to assume. On the other hand an excess demand, if it cannot be shifted (as, we shall see, is the case considered in what follows) will leave the consumers with a stock of money, to carry forward, that they do not wish to hold on to. See Hicks (1965, chapter VII).

[24] Flexible prices, in particular, would make it impossible to bring to light the problem of saturation of demand, which is one of the main determinants of an innovative choice.

us to figure on a breaking of the sequence without being obliged to assume the previous existence of a technological advance that would automatically reduce the process of innovation to the embodiment of such an advance into the prevailing productive structure as is the case in the 'traditional' and, to some extent in the 'new' approach. Money – the increased demand for liquid positions as a result of a modification of long-term expectations which reflects a loss of confidence in the existing model and the search for something new and different – acts as a signal that allows the case for a qualitative change originating *within* the model to acquire analytical relevance through the accumulation of money stocks.[25] The ensuing process of innovation, if an innovative choice is made, will bring about new technology which then appears as the result of such a choice and not as the precondition of it.

This can certainly not happen in a perfect barter economy, where the possibility of a stock disequilibrium is not even likely to arise. The hypothesis of a homogeneous output that can serve as both consumption and investment and, if saved, is immediately and automatically invested in the only possible way (i.e. the hypothesis that the output of a given period is necessarily consumed and invested within it) guarantees in fact a flow equilibrium within each given period.[26] The same can be said, still with reference to a barter economy (or to a virtual barter economy where money is only a unit of account), when the assumption is made of a heterogeneous output which implies the possibility of a mismatch between the structure of supply and the structure of demand that can actually bring about a flow disequilibrium that results in the piling up of disequilibrium stocks. In this case, in fact, the disequilibrium stocks can appear but cannot be signalled as a problem of qualitative change; they must be considered as the result of a wrong structuring of supply and then automatically treated as a problem of quantitative adjustment (more of certain commodities, less of others) (Amendola, 1984b).

[25] The feeling or the perception that something new and different is going or needs to happen could not bring about the breaking of a sequence if the existence of financial reserve assets (money) did not allow it to be translated into an appropriate behaviour.

[26] This is also the case in the Hicksian Neo-Austrian barter model where the flow equilibrium in each period is assured by the rate of starts of new processes of production which is endogenously determined and adjusts so as to absorb all the output not required for consumption or for investment in the old processes of production still being carried on. The suggestion of an exogenously determined rate of starts (Hicks, 1973, pp. 52–4), which might cause a disequilibrium and the accumulation of stocks that would allow an overfunctioning (or an underfunctioning) of the economy over a certain number of periods, is not really convincing and is essentially extraneous to the logic of the model.

In a Neo-Austrian model, in contrast, a stock disequilibrium can only take a monetary form[27] (if money is there to allow it), thus immediately signalling the qualitative nature of the change required. This kind of disequilibrium, on the other hand, carries its effects down the sequence in a different way with respect to the case in which only real stocks are considered, underlining once again the difference between a process interpreted as a qualitative change and a process treated as a quantitative adjustment.

The most immediate consequence of the presence of money stocks in the model, as we have seen, is the possibility of a current final demand falling short of supply, and hence the possibility of the scrapping of the processes of production still in the phase of utilization. The breaking of a sequence in a monetary economy in fact, by modifying the proportion of the existing financial resources actually devoted to production and/or consumption, brings sharply to light the difference between money and capital; it thus makes a 'financial' constraint appear which, together with the already mentioned 'human resource' constraint, is to play a central role in the articulation and in the effective evolution of the process of innovation as a sequential process.

This financial constraint depends on the financial resources (money, in our case) that, at the beginning of each period, are actually devoted to the carrying out of the processes of production and that – as we have already pointed out (pp. 40–1) – may have an internal or an external source. Together with the human resource constraint, and given the expectations, it determines current production and investment decisions (that is, the degree of utilization of the existing productive capacity, and the number and type of the processes of production to be kept going and to be started anew); while, through its effects on the Wages Fund, it also helps to determine current final demand.

The Sequence 'Constraints–Decisions–Constraints'

We now have all the ingredients required for working out a sequence that provides the most adequate framework for the analysis of the process of innovation interpreted as a process of learning that brings about a modification of the environment as a source of technology.

[27] Except in the case that has just been mentioned, put forward by Hicks. It goes without saying that we are referring to the original model with a single homogeneous final product (or with a composite good with fixed proportions, which amounts to the same thing). We are in fact convinced that a Neo-Austrian model cannot be transformed into a true multi-sector model (in which the piling up of real stocks could be considered) without losing its distinctive feature, that is, the full vertical integration of the process of production.

The technical intertemporal complementarity of the process of production and the intertemporal complementarity of the decision process are the two main threads whose interweaving causes the linking up of the successive periods into a particular kind of sequence.

The technical intertemporal complementarity of fully vertically integrated processes of production,[28] on the one hand, creates irreversibilities that establish links between what is required (and what can be done) today (and, to a certain extent, what will be required and what it will be possible to do tomorrow) and what happened yesterday – by determining the spread over time of the human and of the financial requirements of the choices made at each moment, and the lags in the effects of such choices, in a context in which investment of output capacity is not the same thing as investment at cost. Thus the decisions of a given period imply the setting in of irreversible processes that call for other related decisions in the future and, at the same time, place constraints on them.[29]

This is the process through which the rate of starts is made endogenous, thus setting the pace of the 'traverse' in the standard Hicksian analysis.

The link between successive periods, in sequential models not characterized by a technical intertemporal complementarity of the processes of production, is provided on the other hand by the explicit consideration of the expectations, which, when based upon past experience (that is, on what has happened in the preceding period(s)), ensure the connections of plans and decisions over successive periods.[30]

Taking into account both these links, we consider a Neo-Austrian productive context in which the decisions as to current production and investment, given the existing expectations (short- and long-term respectively) depend on the (human and financial) constraints which, at each given moment, limit both the range of the feasible choices and the levels of activity, as only those types of production processes can be carried out for which the required (heterogeneous) labour inputs, and the resources to finance their employment, are available.

---

[28] Emblematically expressed, in the standard Neo-Austrian model, by the vertical structuring of the process into a phase of construction of the productive capacity and, following that, a phase of utilization.

[29] The links thus established have, however, a mechanical character, reflecting the rate at which, for technical reasons, the resources are gradually released from the economy to be employed in the new processes.

[30] As is the case, for example, when non-fulfilment of plans and hence a flow disequilibrium in a given period, implies a revision of the expectations that brings decisions about in the following period aimed at eliminating the disequilibrium.

The constraints, on the other hand, are modified from period to period as the result of decisions taken in the past, but producing their effects in the present through the existing intertemporal complementarities of production (and through the learning process, when this takes place). Thus today's decisions, taken on the basis of today's constraints, go to modify the constraints that will affect tomorrow's decisions – and so on in the sequence 'constraints–decisions–constraints'.

Let us consider first a 'routine' context, defined as a context in which there is no learning. The processes of production carried out and the existing patterns of consumption are the expression of an already established and fully developed technology to which the human resource has already completely adapted itself. Everybody (producers and consumers) is satisfied with what is going on and there is no search for anything new and different; long-term expectations are given and remain in the background; short-term expectations, consistent with the latter, control the decisions taken in each period and have only a quantitative content (*how much* to produce). The only relevant constraint is the financial constraint which affects the level of the activity carried out according to the prevailing technology and which is affected, in turn, by the results of this activity, given the existing technical lags. To sum up, the sequence 'constraints–decisions–constraints' portrays a process that has the nature of a quantitative adjustment, whether the economy is in a state of self-replacement, or growing at a given rate, or trying to get back to a position or to a path from which a mistake in the forming of the expectations or some other temporary disturbance has removed it.

The Learning Process and the Human Resource Constraint

When the effects of a process of learning, besides the already mentioned intertemporal complementarities, are taken into account in the shaping of the pattern of interactions between constraints and decisions, both the nature and the actual articulation of the sequence change.

A learning process, in our interpretation of the process of innovation, is in the first place the result of long-term expectations coming directly into light. It is a modification of the way in which these expectations are formed, in fact, that raises the problem of *how*, and not only of *how much*, to invest, thrusting towards the innovative choices that are the premise of the process of learning that characterizes the process of innovation as a qualitative change.

An innovative choice, on the other hand, means moving in different directions from the usual ones and starting processes of production whose features will only take on precise definition along the way, and to which therefore the existing labour inputs are not yet adjusted. A human resource constraint becomes then effective, limiting the range of feasible choices (that is, the directions along which it is actually possible to move the first and then the successive steps of the process of innovation) and helping at the same time to determine, together with the financial constraint, the levels of the productive activity.

However, the decision to set out along an innovative path, will not only affect the financial constraint in future periods in the usual way (that is, through the already mentioned intertemporal complementarities), but will trigger a process of learning that will bring about a modification of the human resource constraint as well. The carrying out of innovative processes of production, in fact, implies research and experimentation that, while helping to specify the profiles of the processes themselves, at the same time brings about the upgrading of the human resource that has taken part in those processes and causes the appearance of entirely new skills and qualifications that will themselves make it possible to devise and implement new types of processes of production.[31] The human resource constraint thus becomes less stringent as the process of innovation goes on, both in the sense of an enlargement of the range of options that are continuously created anew, and in the sense that some particular bottleneck that might affect the level of productive activity is eliminated.

The interaction between constraints and decisions is thus much more complex on an innovative path than in a routine context; but that is not all. What is more important, and what is really distinctive about a process of learning that brings about new technology and new options through a modification of the features of the human resource, is the relation between the human and financial constraints themselves, which on an innovative path are not independent of one another, but come to be sequentially related.

Learning, we have seen, takes place through the carrying out of innovative processes of production; its intensity depends therefore essentially on the rate of starts of such processes, which at each given moment is limited by the existing financial constraint (together with the human constraint). A less stringent financial constraint thus means, by and large, a more intense learning process.[32] Learning, on the

---

[31] The consumption side of the economy is also involved in this process. New forms of consumption in fact become possible with the defining of new forms of production, making learning in production and learning in consumption appear as two sides of the same process of modification of the environment conceived as a source of technology.

[32] Provided, of course, that the available resources are invested in innovative processes, as we assume to be the case on an innovative path.

other hand, brings about an enrichment and a greater articulation of the human resource, which implies a releasing of the human constraint and the opening of new paths along which to move the subsequent steps. A less stringent financial constraint therefore also means a less stringent human resource constraint; and since the human constraint usually prevails over the financial constraint in the first phases of a process of innovation, its release implies in turn the opportunity to utilize more of the existing, though as yet frozen, financial resources into the processes of production, and so on.

Moreover, as the process of innovation goes on, the accruing of new opportunities at each successive step provides an adequate answer to the search for flexibility that has prompted the producers to start moving on an innovative path, with the effect of reducing their demand for liquid positions and of increasing accordingly the proportion of resources (money) that they are willing to devote to production. In the same way, learning brings about an increasing preference for the new forms of consumption that emerge together with the new forms of production, and hence a greater confidence on the part of consumers in what is going on, with the result of reducing the liquid reserve assets that they are inclined to hold and of increasing their demand for current output. This, in turn, will make it possible to increase the proceeds of the producers – which, together with the change in the distribution between Financial and Industrial circulation, represent the internal source of the resources available for carrying out the processes of production in each given period – thus reinforcing their confidence in the working of the economy and likewise affecting their long-term expectations as well as those of consumers.

The association – which we have seen to hold within each given period between short-term expectations and current production decisions on one side, and between long-term expectations and investment decisions on the other – thus collapses when successive periods linked in a sequence are considered, and that is all the more true if the sequence itself is characterized by a process of learning like the one described.

The interactions between decisions and constraints and between short- and long-term expectations, and the chain of effects that, given these interactions, actually takes place along an innovative path, are however too complex to be clearly worked out in purely logical terms. For that to be possible, we need to 'stylize' the sequential process in such a way as to bring to light the relevant connections that define the analytical articulation of the sequence in time in an explicit form.

This is what the model presented in the next chapter is intended to do. It is meant to show that the interpretation of the process of innovation that we have proposed can be given a rigorous analytical structure, and, what is more relevant, that there are important economic policy implications of the analysis that can thus be sorted out. Such implications will be discussed in detail in chapter 5, both in general and with particular reference to specific propositions for an economic policy conducive to change and innovation.

4
The Model

The Features of the Model

The model presented in this chapter deals with a closed economy in which the agents are divided into two types, producers and consumers, who are considered at an aggregate level. There is also a third agent, say the Government, who remains in the background and whose only task is to inject into the economy the only existing financial asset, 'money'.

We are dealing in fact with a monetary, not a barter economy. The exchanges are made, and the wages are paid, in money terms, and the resources required to carry out the production processes are thus financial resources, not physical output. Borrowing and lending are not considered; producers and consumers are obliged to limit their spending in any period to what is made available to them in the same period as a result of their participation in production, plus the stocks of money that they have voluntarily or involuntarily accumulated in the past. The only exception is represented by the exogenous inflow of money – expressed in the model by the value taken by the exogenous variable ΔM – decided upon and supplied by the Government to the producers at the beginning of each period. ΔM, which measures the indebtedness of the producers, is the main instrument available to the Government to pursue its policy (if any).

Production is carried on in a Neo-Austrian context, in which capital goods are internal to each process, and hence are implied but not explicitly shown. The production process is thus portrayed as a scheme for converting over time a sequence of heterogeneous and specific

53

labour inputs into a sequence of final output. The specific character of the labour inputs is the result of the skills and qualifications acquired through a learning process by human resources involved in innovative processes of production, that is, in processes characterized by research and experimentation. Conversely, participation in production processes carried on according to a given, and fully developed, technology, is not characterized by learning and thus does not imply (further) specification of human resources.

These extremely simplifying assumptions are made in order to concentrate our attention, and to throw light, on the relevant aspects of a process of innovation interpreted as a process of modification of the environment conceived as a source of technology. In particular, the model proposed is intended to portray the evolution of the economy through a sequence of periods, defined as 'decision' periods and made to coincide with the elementary periods of the production process. The 'decision' or 'single-period' analysis shows how certain decisions are taken at the beginning of each period, and how they lead to certain results at the end of it. The technical intertemporal complementarity of production processes fully integrated vertically, and the intertemporal complementarity of the decision process, link up the successive periods in a sequence. The sequences that actually take place in the economy, as we shall see, depend strictly on the presence of money and on the role that it plays in the model.

The relations between the relevant magnitudes of the economy reflect the sequential structure of the model both within each period and between successive periods.

At the beginning of each period producers can count on financial resources ('money') that have an external source – the Government – and/or an internal source: the proceeds of the sales of the final output of the preceding period(s). The available resources (a part of which is kept aside by the producers for their own consumption) are fully or partially channelled into production according to the value taken by a parameter ϱ which depends on the state of long-term expectations that reflect the degree of confidence of the producers in the way in which the economy is actually functioning. The resources actually devoted to production make up the Wages Fund which, together with the existing human resources, sets a constraint on the number and type of production processes that can be effectively carried on. Given these constraints, decisions on current production and on investment depend on the short-term expectations (which concern the demand for final output expected in the period and determine both the amount and the composition of current production, that is, the degree of utilization of the productive capacity inherited from the past) and on the long-term expectations of the producers respectively.

On the other hand, the final output obtained at the end of the period is matched by a demand whose amount and composition, given the consumers' preference system, depends on the fraction of the available financial resources (that is, the wages advanced by the producers to the workers who have taken part in the processes of production, plus the money kept by the producers for their own consumption, say Q) actually devoted to consumption. This fraction is determined in the model by a parameter σ that depends on the consumers' long-term expectations which, in the same way as the producers' long-term expectations, reflect the degree of confidence of the consumers in the way in which the economy is actually functioning.

As we have already pointed out, additional resources for production and/or consumption in each period can come from the money stocks (if any) voluntarily accumulated as liquid assets in previous periods, as a result of ϱ and/or σ having taken values lower than unity in the same periods. In this case ϱ and/or σ can take a value greater than unity in the period considered, thus making it possible to reabsorb these stocks into production.

However, producers' and consumers' decisions are not necessarily mutually consistent. A flow disequilibrium can thus appear at a given moment, which cannot be taken care of by price changes (prices and wages are assumed to remain unchanged throughout each given period) and will result in the piling up of stocks of final output (if there is an excess supply) or of stocks of money that the consumers do not intend to hold on to (if there are shortages of supply). These stocks are automatically shifted to the following period, and this helps to carry the disequilibrium down the sequence.

In this chapter we examine in particular the working of the model and its solution in any given period. This solution depends on the values taken by the exogenous variables and by the parameters of the model, and, as we shall see, differs according to whether the financial constraint or the human resource constraint prevails.

Modifications in the producers' and/or consumers' behaviour that call for qualitative changes in the functioning of the economy are expressed in the model by changes in the value of the parameters ϱ and/or σ, which necessarily result in a disequilibrium. The response to such a disequilibrium can either be a quantitative adjustment that consists in a simple revision of the final output and investment targets of production processes that continue to be carried on according to the already established technology, or an innovative choice that will bring about a process of creation of technology through a modification of the environment.

The model has been submitted to some numerical experiments in order to throw light on the patterns of evolution of the relevant magnitudes of the economy (output, investment, employment, indebtedness and so on) both in the case of a routine choice and of an innovative choice, under alternative hypotheses on the behaviour of ΔM, featuring different policies followed. This has been done not so much because the formal study of the model appears to be too difficult as because we are interested in the exploration of processes whose possible convergence to given points or paths – as has already been pointed out in chapter 1 – has no analytical relevance in the perspective adopted. The results obtained are discussed in chapter 5, with particular reference to the conditions required for an innovative choice (which, as we shall see, is the only appropriate answer when the case for a qualitative change arises) to be viable, and to the economic policy implications of the analysis carried out.

The Process of Production

The process of production, defined as a scheme for converting heterogeneous labour inputs into final outputs, goes through a sequence of periods that, following Hicks's device (Hicks, 1970, 1973), are grouped in two phases: the phase of construction and, subsequently, the phase of utilization of (a unit of) productive capacity. This is a useful device for exhibiting some important consequences of the technical intertemporal complementarity of the process of production, but the sequence of elementary periods – each characterized by the application of fresh labour to production and, when the process has reached the phase of utilization, also by the carrying out of a round of current final production – must be taken as a whole to represent a process of production fully integrated vertically. At the same time the distinction between the phase of construction and the phase of utilization must not be interpreted as marking necessarily the appearance of physical equipment; it certainly does no harm to keep thinking traditionally in terms of plants and 'machines', but what is really meant here by the phase of construction is more generally the preliminary setting of the stage for actual final production, whether it involves physical capital goods or not.

The processes of production carried out in the economy are defined as either 'routine' (x) processes, or ' innovative' (y) processes.

Routine Processes   Routine processes of production are the expression of an already established and fully developed (in the sense that no substantial improvements or refinements are expected) technology which reflects the skills of existing human resources. To such a technology the productive capacity of the economy and the structure of its heterogeneous labour force, which is its meaningful expression in a model where the only inputs shown are labour inputs, are assumed to be already perfectly adjusted.

An elementary routine process – which makes it possible to obtain from a unit of productive capacity a composite basket of storable commodities and/or services whose characteristics are also fully defined and perfectly known to the consumers, who have already learnt the forms of consumption implied by the prevailing technology and adjusted their preferences accordingly – is described by

1 a matrix $\boxed{A}$ whose elements α_{hk} represent the quantities of the different types of labour ($h = 1,2,...,s$) required by the process in the different periods of its life ($k = 1,2,...,m,m+1,...,m+M$), where m and M are the length of the construction phase and the length of the utilization phase respectively; and which can be partitioned into the submatrices $\boxed{A}^c$ and $\boxed{A}^u$ referring to the same phases;
2 a vector α_0 whose elements represent the heterogeneous labour requirements for starting the process of production;
3 the row vector of final outputs c, whose elements refer to the amounts of final output obtained in the different periods of the phase of utilization of the process; that is

$$\boxed{A} = [\ \boxed{A}^c \ \ \boxed{A}^u\ ]$$

$$\boxed{A}^c = [\alpha_{hk}^c] \qquad h = 1,2,...,s,\ k = 1,2,...,m$$

$$\boxed{A}^u = [\alpha_{hk}^u] \qquad h = 1,2,...,s,\ k = m+1,...,m+M$$

$$\alpha_0' = [\alpha_{10},\alpha_{20},...,\alpha_{s0}]$$

$$c' = [c_{m+1},c_{m+2},...,c_{m+M}]$$

where the prime denotes transposition.

Innovative processes Conversely an innovative process of production is not the expression of an already established technology in a perfectly adjusted productive context, but a moment in a transformation under way: an intermediate step in a process of research and experimentation that will have different developments according to the different paths opened up by different successive steps, and to the different choices made in a related sequence.

The profile and the characteristics of the process of production will thus change at each successive step, this change being the expression of the ongoing technological and productive transformation. The elements of the matrix $\mathbf{B}$ and of the vector β_0, which are the exact replicas for an innovative process of production of the matrix $\mathbf{A}$ and of the vector α_0 for a routine process, must not therefore be considered as technical coefficients in the same way as the elements of the latter, but should be seen as the initial requirements for starting and carrying on an innovative process – requirements that will change together with the process of production itself as this gets more and more fully specified. Only the quantitative aspect of this change, however, is captured here – through the assumption that, on the whole, the labour requirements will decrease in time. Time is taken as starting from the moment when each particular type of labour is first applied, at a rate that, however, does not depend on the mere passage of time, but on the workforce's acquaintance with the new productive problems. The latter is a function of the number of innovative production processes carried out from the moment 0, when an innovative choice was first made, up to the particular period considered; that is

$$\mathbf{B}(t) = [\ \mathbf{B}^c(t)\ \ \mathbf{B}^u(t)\,]$$

$$\mathbf{B}^c(t) = [\beta_{hk}^c(t)] \qquad h = 1,2,\ldots,s,\ \ k = 1,2,\ldots,n$$

$$\mathbf{B}^u(t) = [\beta_{hk}^u(t)] \qquad h = 1,2,\ldots,s,\ \ k = n+1,\ldots,n+N$$

$$\beta_0'(t) = [\beta_{10}(t),\ \beta_{20}(t),\ldots,\beta_{s0}(t)]$$

with

$$\beta_{h0}(t) = f_0\left(\sum_{T=0}^{t-1} y_0(T)\right) \qquad \beta_{h0}(0) = \bar{\beta}_{h0},\ f_0' < 0, \forall h$$

$$\beta_{h1}(t) = f_1\left(\sum_{T=1}^{t-1} y_1(T)\right) \qquad \beta_{h1}(1) = \bar{\beta}_{h1},\ f_1' < 0, \forall h$$

$$\cdot \qquad \cdot \qquad \cdot \qquad \qquad \cdot \qquad \cdot \qquad \cdot$$
$$\cdot \qquad \cdot \qquad \cdot \qquad \qquad \cdot \qquad \cdot \qquad \cdot$$
$$\cdot \qquad \cdot \qquad \cdot \qquad \qquad \cdot \qquad \cdot \qquad \cdot$$

$$\beta_{h\,n+N}(t) = f_{n+N}\left(\sum_{T=n+N}^{t-1} y_{n+N}(T)\right)$$

$$\beta_{h\,n+N}(n+N) = \bar{\beta}_{h\,n+N},\ f_{n+N}' < 0, \forall h$$

where $f_0', f_1', f_2',\ldots,f_{n+N}'$ are the first derivatives of the functions.

In the same way the elements of the final-output vector

$$d'(t) = [d_{n+1}(t), \ d_{n+2}(t),...,d_{n+N}(t)]$$

refer to amounts of storable commodities and/or services whose composition and features will change together with the profile of the process of production, as the process of innovation proceeds.

The Creation of Technology as a Learning Process

A process of creation of new productive options, associated with the transformation of the environment as expressed mainly by a modification of human resources, is, on the other hand, fuelled by the learning process that sets in as the result of an innovative choice and of the carrying on of innovative processes of production.

Learning consists in a widening of the available information in the uncertain context of the search for new solutions to (new) production problems, and is the result of the greater familiarity with those problems acquired while taking part in innovative processes of production and contributing to their specification. It concerns both the production and the consumption side of the economy, and takes the form of the acquisition of higher and sometimes entirely new skills, and, as we shall see, of a modification in the consumer's preference system.

The upgrading and the greater articulation of human resources, obtained in the carrying out of specific (innovative) processes of production, suggests new approaches and makes it possible in turn to define new types of processes of production. The appearance of new and different skills, in fact, implies thinking and organizing production (and consumption) in altogether original ways.

Technology, thus, no longer appears as a specific way (with its own physical counterpart) to solve a given problem, but as an environment characterized by human resources capable of devising, and implementing, different solutions to different problems. While in the traditional approach the process of innovation is seen as the adjustment of the economy and of its resources to a given (superior) technique, here it is a modification of the existing resources – namely the appearance of entirely new skills, and hence of new specific labour inputs – which will change the environment itself as a source of technology.

We assume therefore that the productive capacity of the economy reflects the way in which the processes of production are actually articulated, and that the latter are shaped according to the particular features and the structure of the existing labour resource.

A labour availability vector can then be written at each time t

$$\boldsymbol{L}_s^{S'}(t) = [l_1^S(t),\ l_2^S(t),...,l_s^S(t)]$$

whose elements represent the different skills of the heterogeneous labour resource, and where the term 'skill' must be taken in a wider sense as referring to all the relevant characteristics (qualification, mobility, organizational capabilities, institutional features, . . .) of this resource in that it is the expression of a given environment. We assume further that these skills are ranked in an increasing order, the lower-indexed elements corresponding to the lower skills, and the higher-indexed elements to the higher skills (that is, the skills that reflect more experience and that have therefore more recently appeared), and that those who have various skills and can thus perform different jobs are classified under the item corresponding to their higher skill at that moment.

This labour availability vector gets modified as time goes by both in a routine context and in an innovative context, but in different ways.

When only routine processes of production are carried out according to a given technology, and there is no learning, only external inflows (outflows) reflecting demographic and educational factors and affecting the size of the different elements of the vector, must be considered. This can affect all the elements of the vector in the same proportion or each of them in a different way, thus altering the structure of the vector itself to a greater or lesser extent. What we are dealing with, in any case, are changes in absolute and/or relative size: that is, a quantitative phenomenon.

Where innovative processes are considered, in contrast, the process of learning must be taken into account, besides the demographic and educational factors already mentioned. The effect of learning on the labour resource is twofold. On one side, as more and more innovative processes are being carried on and the workers employed in them get acquainted with the new production problems, they gradually acquire higher skills within the range of those associated with the existing profiles of the production process. We have thus an internal upgrading process resulting in an outflow from each element of the vector and in a corresponding inflow in the next-higher-indexed element: the net additional balance represents the change in the size of each element as a result of this process. Once again, we are considering only changes in size, of a quantitative nature.

Learning, however, means not only this, but also the appearance of entirely new skills, that is of ones that are not associated with the existing profiles of the production process, but that will themselves bring about new profiles. Thus as time goes by and more innovative processes are being carried on, new elements representing new skills are added to the existing ones, so not only the size of the labour availability vector's elements but its very dimension is modified. Changes in the labour resource, in other words, are not only of a quantitative kind, but also have a qualitative character.

A different labour availability vector will thus define the structure of the labour resource at the beginning of each successive period (assuming that the elementary period of production is the span of time during which the changes take place discontinuously). Each successive vector will generally have elements of a greater size and, when learning is considered, also a greater dimension (for simplicity, only one element is added in each period to the vector), that is:

$$L_s^{S'}(t) = [l_1^S(t),\ l_2^S(t),...,\ l_s^S(t)]$$

$$L_{s+1}^{S'}(t+1) = [l_1^S(t+1),l_2^S(t+1),...,l_s^S(t+1),l_{s+1}^S(t+1)]$$

$$\cdots$$

Each element of the availability vector, in each given period, is the result of demographic and educational factors, and, when it takes place, of the learning process, that is:

$$l_h^S(t+1) = l_h^S(t) + a_h l_h^S(t) - b_h l_h^{DI}(t) + b_{h-1} l_{h-1}^{DI}(t).$$

a_h is the proportion of workers of skill h accruing in the period $t+1$, and depends on the hypotheses made as to demographic rules and educational policies. Thus a_h can be a scalar or a more or less complicated function, which can be different for each of the various elements of the vector or the same for all of them. b_h is the proportion of workers of skill h employed in the innovative processes in the period t, $l_h^{DI}(t)$, acquiring a higher skill in the period $t+1$, and hence moving from the $h-$ element to the $(h+1)$-element of the vector $L_{s+1}^S(t+1)$. b_{h-1} is the proportion of workers of skill h-1 employed in the innovative processes in the period t, $l_{h-1}^{DI}(t)$, acquiring a higher skill in the period $t+1$, and hence moving from the $(h-1)$-element to the h-element of the vector $L_{s+1}^S(t+1)$. Also,

$$b_h(t) = b_h \left(\sum_{T=0}^{t} y(T) \right) \qquad 0 = \text{the moment when the}$$

innovative choice was first made and hence

$$b_h(t) - b_h(t-1) > 0 \qquad \text{assuming that } \sum_{T=0}^{t} y(T) > \sum_{T=0}^{t-1} y(T)$$

that is, assuming that it is not the mere passage of time that matters for learning to take place, but the fact that as time goes by more innovative processes are carried out, so the internal upgrading due to learning is an increasing function of the number of those processes carried out up to the particular period considered.

The labour availability vector in the same period can then be written as

$$
L_{s+1}^{S}(t+1) =
\begin{bmatrix} L_s^S(t) \\ \\ \cdot \cdot \cdot \\ 0 \end{bmatrix}
+
\begin{bmatrix} a_1 \ldots \ldots 0 \\ \ldots \ldots \ldots \\ \cdot \quad \cdot \quad \cdot \\ \cdot \quad \cdot \quad \cdot \\ \cdot \quad .a_s \quad \cdot \\ 0 \quad . \quad 0 \end{bmatrix}
\begin{bmatrix} L_s^S(t) \\ \\ \cdot \cdot \cdot \\ 0 \end{bmatrix}
-
$$

$$
\begin{bmatrix} b_1 \ldots . 0 \\ \cdot \quad \cdot \quad \cdot \\ \cdot \quad \cdot \quad \cdot \\ \cdot \quad \cdot \quad \cdot \\ \cdot \quad .b_s \quad \cdot \\ 0 \ldots . 0 \end{bmatrix}
\begin{bmatrix} L_s^{DI}(t) \\ \\ \cdot \cdot \cdot \\ 0 \end{bmatrix}
+
\begin{bmatrix} 0 \ldots . 0 \\ \cdot \quad b_1 \quad . \quad . \\ \cdot \quad \cdot \quad \cdot \\ \ldots \ldots . \\ \ldots \ldots . \\ 0 \quad . \quad . \quad b_s \end{bmatrix}
\begin{bmatrix} 0 \\ \cdot \cdot \cdot \\ \\ L_s^{DI}(t) \end{bmatrix}
$$

where $L_s^{DI}(t)$ is the vector whose elements represent the various types of labour employed in the innovative processes in period t, and the last two terms on the RHS – which reflect the effects of the internal upgrading due to the learning process and hence show the appearance of an additional element with respect to the preceding period – must be considered only when innovative processes are taken into account and the learning process operates over the vector $L_s^{DI}(t)$.

The Human Resource Constraint

Given the range of the processes of production that the skills of the existing human resource make it possible to devise, the labour availability vector sets a constraint both on the type and the number of the processes that can actually be carried on in each given period, as only those processes can be started and/or kept alive (and in the amounts) for which the required labour inputs are available in the right proportions. Given in fact a labour demand vector resulting from the producers' decisions on routine and on innovative processes

$$
L_s^{D'}(t) = [I_1^D(t), I_2^D(t), \ldots, I_s^D(t)] = L_s^{DR'}(t) + L_s^{DI'}(t)
$$

and assuming, as we are, that for the routine processes the structure of the labour supply is perfectly adjusted to the structure of the demand (that is, $L_s^{SR}(t) = L_s^{DR}(t)$) the vector $L_s^{DI}(t)$ of the labour requirements for the innovative processes must be confronted with the availability vector $L_s^{SI}(t) = L_s^{S}(t) - L_s^{DR}(t)$, and if the structure of this residual supply does not coincide with the structure of demand, and

$$I_h^{SI}(t) < I_h^{DI}(t) \qquad \text{for some } h = 1,2,\dots,s$$

we shall have a 'constrained' labour demand vector for innovative processes:

$$\hat{L}_s^{DI}(t) = \left(\min_h \frac{I_h^{SI}(t)}{I_h^{DI}(t)} \right) L_s^{DI}(t)$$

which limits the number and type of innovative processes that can actually be kept going – since each type of process requires a particular combination of skills in given proportions – so the aggregate labour demand vector becomes

$$\hat{L}_s^{D}(t) = L_s^{DR}(t) + \hat{L}_s^{DI}(t)$$

(in fact, we have various labour demand vectors for innovative processes, which reflect the age structure of those innovative processes that, together with the routine ones, actually make up the productive capacity of the economy).

However, this constraint is not strictly rigid; it is modified by the process of learning in the longer run, as shown in particular in the previous section, but can also be made less stringent in the very short run. The hypothesis that those who are able to perform various jobs are classified under the item corresponding to their higher skill, in fact, gives 'malleability' to the labour availability vector, in the sense that there is a certain flexibility towards the lower layers. Thus in each given period, if possible and necessary, the vector itself can be restructured, by moving the workers from the higher- to the lower-skill elements, to suit better the labour demand vector.

On the other hand a less stringent labour constraint implies both an enlargement of the range of feasible processes (including already known types of processes of production that were not feasible in the past because of the total or partial lack of some kind of labour that is now available, and entirely new processes defined by the appearance of entirely new skills), and an increase in the absolute number of processes of any kind that can actually be started and carried out.

However, that is not all. Learning, in fact, has two faces. In our interpretation of technology and of its changes, as we have already pointed out, the production and the consumption side are strictly related. In this context the enrichment and the greater articulation of human resources, as a result of the greater familiarity with the new production problems as we proceed along an innovative path, then also imply new forms of consumption which, in turn, help to define new products and to articulate new processes of production.

Thus as the process of innovation goes on the main features of consumption, expressed in the model by the consumers' preference system, get transformed along with the technology. We have tried to render this process by assuming, in particular (see below pp. 74–5), that the degree of preference for the output of the innovative processes increases as more of these processes are carried out, and as the consumers get better acquainted with products whose modifications they help to define.

The Working of the Model

The Economy

Productive Capacity   The state of the economy at any given time reflects the decisions taken in a related sequence up to the moment considered, given the technical intertemporal complementarity of production.

In particular the age structure of the stock of processes that represents the productive capacity of the economy is the result of the successive decisions of the producers regarding current production and investment, which may imply the scrapping of the processes before the end of their physical life. Therefore, given the number and type of the processes of production of different ages inherited from the past to which labour can still be applied, the actual configuration of the productive capacity in each given period will depend on the number and type of the processes scrapped (to which labour is no longer effectively applied) and on the number and type of processes that are started from scratch (rate of starts) in the same period.

Let $x^c(t+1)$ and $x^u(t+1)$ be the vectors whose elements represent the number of the routine processes, in the periods 1 to m of the construction phase and in the periods $m+1$ to M of the utilization phase respectively, to which labour is applied at time $t+1$, and $y^c(t+1)$ and $y^u(t+1)$ the corresponding vectors for the innovative processes; and let $u^c(t+1)$, $u^u(t+1)$ and $v^c(t+1)$, $v^u(t+1)$ be the vectors of the routine and of the innovative processes respectively, in the construction and in the utilization phase, whose elements represent the processes scrapped at time $t+1$.

These vectors are defined as follows. Firstly

$$x^{c'}(t+1) = [x_1^c(t+1),x_2^c(t+1),\ldots,x_m^c(t+1)]$$

$$x^{u'}(t+1) = [x_{m+1}^u(t+1),x_{m+2}^u(t+1),\ldots,x_{m+M}^u(t+1)]$$

with

$$x'(t+1) = [x^{c'}(t+1),x^{u'}(t+1)].$$

Secondly,

$$y^{c'}(t+1) = [y_1^c(t+1),y_2^c(t+1),\ldots,y_n^c(t+1)]$$

$$y^{u'}(t+1) = [y_{n+1}^u(t+1),y_{n+2}^u(t+1),\ldots,y_{n+N}^u(t+1)]$$

with

$$y'(t+1) = [y^{c'}(t+1),y^{u'}(t+1)].$$

Thirdly,

$$u^{c'}(t+1) = [u_1^c(t+1),u_2^c(t+1),\ldots,u_m^c(t+1)]$$

$$u^{u'}(t+1) = [u_{m+1}^u(t+1),u_{m+2}^u(t+1),\ldots,u_{m+M}^u(t+1)]$$

with

$$u'(t+1) = [u^{c'}(t+1),u^{u'}(t+1)].$$

Lastly,

$$v^{c'}(t+1) = [v_1^c(t+1),v_2^c(t+1),\ldots,v_n^c(t+1)]$$

$$v^{u\prime}(t+1) = [v_{n+1}^u(t+1),v_{n+2}^u(t+1),\ldots,v_{n+N}^u(t+1)]$$

with

$$v'(t+1) = [v^{c'}(t+1),v^{u'}(t+1)].$$

The rates of starts of the routine and innovative processes at time $t+1$, on the other hand, are defined respectively by the scalars $x_0(t+1)$ and $y_0(t+1)$, such that

$$x_0(t+1) \geqslant 0 \qquad y_0(t+1) \geqslant 0.$$

The values actually taken by the above-mentioned vectors and scalars, and hence the particular configuration of the productive capacity of the economy, will depend on the particular solution of the model in each successive period (see below pp. 75–8 ff).

Macroeconomic Magnitudes   We now list the relevant magnitudes of the economy.

1   Total output, reckoned as the money value of aggregate final output in each given period (e.g. in period $t+1$), and given by

$$P(t+1) = p_R(t+1)(c'x^u(t+1) + R_s(t)) + p_I(t+1)(d'(t+1)y^u(t+1) + I_s(t))$$

where $p_R(t+1)$ is the price of the output of the routine processes in terms of money, $p_I(t+1)$ the price of the output of the innovative processes, likewise in terms of money, $c'x^u(t+1)$ and $d'(t+1)y^u(t+1)$ the current output of the routine and of the innovative processes respectively, and $R_s(t)$ and $I_s(t)$ the stocks of final output of the two types of processes involuntarily accumulated in the previous period, that is, $R_s(t) = R(t) - R^*(t)$ when $R(t) > R^*(t)$, and $I_s(t) = I(t) - I^*(t)$ when $I(t) > I^*(t)$.

Writing

$$c'x^u(t+1) + R_s(t) = R(t+1)$$

and

$$d'(t+1)y^u(t+1) + I_s(t) = I(t+1)$$

we have

$$P(t+1) = p_R(t+1)R(t+1) + p_I(t+1)I(t+1).$$

2   The Wages Fund; that is, the amount of financial resources ('money') required by the labour applied to start and to continue carrying on the routine and/or the innovative processes of production, both in the construction and in the utilization phase, in the current period, given by

$$W(t+1) = w'(t+1)(\boxed{\text{A}}\ x(t+1) +$$
$$\boxed{\text{B}}\ (t+1)y(t+1) + \alpha_0 x_0(t+1) + \beta_0(t+1)y_0(t+1))$$

where

$$w'(t+1) = [w_1(t+1), w_2(t+1), \dots, w_{s+1}(t+1)]$$

is the vector of the money wage rates, exogenously determined, which correspond to the different types of labour employed in the processes of production.

3 The money value of the aggregate demand for final output at the ruling prices

$$P^*(t+1) = p_R(t+1)R^*(t+1) + p_I(t+1)I^*(t+1)$$

where R^* and I^* are the amounts of final output of the routine and of the innovative processes respectively demanded. What determines P^* will be specified later.

4 The producers' financial resources available for final demand (consumption out of profits)

$$Q(t+1)$$

which is one of the components of $P^*(t+1)$, and which is exogenously determined.

5 The exogenously determined inflow of 'money'

$$\Delta M(t+1)$$

which is the difference between loans and repayments.

Decision Processes

The relations between the relevant magnitudes of the economy reflect the sequential structure of the model, both within each period and between successive periods.

Let us first consider what happens in the 'single' period, which is made here to coincide with the elementary period of the process of production. In each period decisions are taken, concerning both the production and the consumption side of the economy, that lead to results not only in the period itself but also in future periods. Production and consumption decisions are assumed to depend: (a) on the existing (financial and human resource) constraints, which limit the range of the feasible choices and the level of the productive activity, and (b) on producers' and consumers' expectations.

The single or 'decision' period is thus defined as such that changes in expectations and constraints do not occur within it, but only at the junction between one period and the next. It is further assumed that prices and wages, set by the producers at the beginning of each period, are maintained unchanged throughout it whatever the market conditions may be.

Long-term expectations – as we have seen in chapter 3 – determine the proportions of the existing financial resources ('money') that the producers on one side and the consumers on the other actually devote to production and consumption.

The resources available to the producers for financing the processes of production (and their own consumption), include (1) the proceeds of the sales of the previous period

$$M(t) = P^*(t) - p_R(t)S_R(t) - p_I(t)S_I(t)$$

i.e. the money value of the aggregate demand for final output in that period, minus the money value of the fractions of such demand – that is $p_R(t)S_R(t)$ and $p_I(t)S_I(t)$ for the routine and the innovative output respectively, where $S_R(t) = R^*(t) - R(t)$ and $S_I(t) = I^*(t) - I(t)$, when $R^*(t) > R(t)$ and $I^*(t) > I(t)$ – which have not been satisfied, resulting in money stocks that the consumers do not intend to keep (we assume that these money stocks are automatically shifted, and go to increase the consumers' money income, and their demand, in the following period); and (2) the exogenous inflow of 'money' $\Delta M(t+1)$.

The Wages Fund available at the beginning of each period for carrying on the processes of production can be written as

$$W(t+1) = \varrho(t+1) \ (M(t) + \Delta M(t+1) - Q(t+1))$$

where $\varrho \ (t+1)$ is a parameter that translates the state of the producers' long-term expectations.

The money value of the aggregate demand for final output, i.e. the fraction of the money income available in each period actually devoted to the purchase of final output in the same period, will on the other hand be

$$P^*(t+1) = \sigma(t+1) \ (W(t+1) + Q(t+1) + p_R(t)S_R(t) + p_I(t)S_I(t))$$

where $\sigma \ (t+1)$ is a parameter that reflects the consumers' long-term expectations.

First consider the producers. At the beginning of each period they must take two kinds of decisions: (1) decisions on current output (and on its price); (2) decisions on investment.

Given the financial constraint as it has been just defined, and given the human resource constraint, these decisions will reflect the short- and the long-term expectations of the producers respectively.

Current Production and Prices Decisions as to current final production are decisions as to the rate of utilization of the existing productive capacity, which in turn is the result of decisions taken in the past, given the technical intertemporal complementarities of the processes of production. They concern the amount of production and, when more than one type of process of production is being carried on at the same time, also its composition: they are therefore decisions as to the number and the type of processes, already in the utilization phase, to which labour must be applied in the current period.

First of all, it must be pointed out that, to arrive at such decisions, the producers must take into account also the shifts in demand and/or the stocks of final output involuntarily accumulated in the previous period, as a result of disequilibria (if any) in the 'routine' market and/or the 'innovative' market, which in the model cannot be taken care of by price changes as we have assumed that such changes do not take place within each given period.

We assume then, in particular, that, at the beginning of each period, the money value of total final production is determined on the basis of the expected money value of aggregate final demand in the same period:

$$P(t+1) = eP^*(t+1)$$

and that these short-term expectations extending over the single period are obtained in the simplest way, by considering a value of the final demand resulting from a growth rate equal to the one realized in the previous period and then adding to it the value of the demand that could not be satisfied in that period, and that is therefore expected to be shifted to the current period:

$$eP^*(t+1) = (1 + g(t))P^*(t) + p_R(t)S_R(t) + p_I(t)S_I(t)$$

where

$$g(t) = (P^*(t) - P^*(t-1))/P^*(t-1).$$

When only routine processes are carried out, and if the economy is in a steady state, the price p_R of the only existing type of output can be set equal to one in terms of money and kept constant over the equilibrium sequence, so

$$eP^*(t+1) = (1+g)P^*(t) \qquad \forall\, t$$

and the decisions as to the amount of output and its value coincide.

If, on the contrary, there is a break in the sequence, the resulting flow disequilibrium will bring about a change in the price of the final output in the following period: the producers will in fact charge a higher price to cover a shift of demand in case of excess demand, or a lower price if they wish an unwanted stock to be reabsorbed in case of excess supply; so we shall have

$$p_R(0) = 1$$

and

$$\frac{p_R(t+1) - p_R(t)}{p_R(t)} = K_R(t+1)\,\frac{R^*(t) - R(t)}{R(t)}$$

where $K_R(t+1)$ is a price reaction coefficient which we can assume to be different, for example, according to whether R^* is higher or lower than R.

When both routine and innovative processes of production are carried out at the same time, the producers must determine the composition of final output (and the money prices) together with its aggregate value.

With regard to this problem we must distinguish the period in which the output of the innovative processes first appears on the market from the successive periods.

As to the first period we assume that:
1 given the price of the routine output already on the market, determined as shown above;
2 given the budget constraint

$$P(t+1) = p_R(t+1)R(t+1) + p_I(t+1)I(t+1)$$

where $P(t+1)$ is determined by the short-term expectations of the producers as shown above; and
3 assuming that all the output of the innovative processes brought to the market is expected to be absorbed whatever the price charged

for it (an assumption whose relevance in this model will be discussed in what follows), so that $I(t+1)$ is determined by the inherited productive capacity, that is, by the decisions on the innovative processes taken in the past, given the technical intertemporal complementarity of production;

then the composition of the final output (and the price of the innovative output associated with it) chosen by the producers will depend on a utility function that the producers themselves expect to be the consumers' utility function in the current period. Let this function be

$$eU^* = U = R^\gamma I^\delta \qquad \text{with } \gamma + \delta = 1.$$

$R(t+1)$ and $p_1(t+1)$ will then be such that

$$p_R(t+1)R(t+1) = \gamma(t+1)P(t+1)$$

$$p_1(t+1)I(t+1) = \delta(t+1)P(t+1)$$

where $R(t+1)$ and $I(t+1)$, as above determined, are the quantities supplied by the producers at the end of the period, while the quantities actually produced will be

$$R_e(t+1) = \min\{R(t+1) - R_s(t); CR(t+1)\}$$

$$I_e(t+1) = \min\{I(t+1) - I_s(t); CI(t+1)\}$$

where CR and CI – which will be defined explicitly in the next section – are the productive capacities of the routine and of the innovative output respectively, inherited from the past.

As to the following periods, (a) we assume that the money price of the output of innovative processes, p_1, will change in response to shifts in demand or to the carrying of stocks from one period to another, in the same way as the price of the routine output p_R (whose determination and changes we have already discussed); that is

$$\frac{p_1(t+1) - p_1(t)}{p_1(t)} = K_1(t+1)\frac{I^*(t) - I(t)}{I(t)}$$

where $K_1(t+1)$ is a price reaction coefficient (which, in the same way as for K_R, we can assume to differ according to whether I^* is higher or lower than I), and (b) we also assume that

$$\delta(t+1) = e\delta^*(t+1) = \delta^*(t)$$

that is, that the consumers' preference for the output of the innovative processes expected by the producers in each given period is equal to the preference the consumers have actually shown in the preceding period. Given the prices p_I and p_R and given P, the model then makes it possible to determine I (which is no longer equal to capacity output) together with R.

Investment   Investment decisions concern: (1) the number and type of the processes of production whose construction, already started, must be continued in the current period, and (2) the rate of starts, i.e. the number and type of the processes that must be started from scratch. These decisions, on *how* and *how much* to invest, are determined simultaneously by the existing financial and human resource constraints and by the long-term expectations of the producers, given a flexibility criterion that we have discussed at length in chapter 3.

We have seen in fact that the problem of *how* to invest first arises when the breaking of a sequence (which brings about a disequilibrium of a qualitative character on a routine path) puts the case for an innovative choice. We have also seen that an increased demand for liquid positions – the first reaction to the breaking of a sequence – is not the appropriate answer to the search for flexibility on the part of the producers when their prospect of learning concerns a learning process that can only come about (and be appropriated) by actually starting off down an innovative path. This, on the other hand, implies an active interpretation of the criterion of flexibility: 'to *increase* the options for the future', which, we repeat, can only be the result of an innovative choice.

In this context the problem of the choice itself (i.e. of *how* to invest) shades off, giving way to the problem of the viability of an innovative choice, which – as we shall see in particular in the next chapter – depends mainly on the rate of starts of innovative processes which determines first the setting in and then the intensity of the process of learning. The problem then becomes relevant of *how much* – and in particular *when* in the course of the process of innovation – to invest in innovative processes of production.

Long-term expectations play an important role in this problem, namely by determining the proportion of the available financial resources channelled into the Wages Fund through the value assumed by the parameter ϱ. We have seen that it is a modification of the long-term expectations of the producers, due to a loss of confidence in the working of the economy, that brings about a decrease in the value of the parameter ϱ, the breaking of a sequence and – when rightly interpreted as the signal of a qualitative change – the appearance of the case for an innovative choice. We assume now that as more and more innovative processes are carried out, intensifying the process of learning and thus bringing about new options that represent an answer to their search for flexibility, the confidence of the producers in the new way of working of the economy will increase, and this will bring about an increase in the value of the parameter ϱ as a result of a modification in their long-term expectations; that is

$$\varrho(t+1) = \varrho \left(\sum_{T=n+1}^{t} I(T) \right) \qquad \Delta\varrho(t+1) > 0.$$

Consumption Decisions on aggregate final demand and – when more than one type of output is available on the market – on its composition, reflect the consumers' preference system (we assume for simplicity, as we have already pointed out, that the consumption out of profits Q is exogenously determined).

In an uncertain and irreversible context, the decision process is represented by a related sequence of choices. To describe it, we do not use an intertemporal utility function, as this would prevent us from taking into account the consideration that more information – both as to the arguments entering the preference system and as to their characteristics – can accrue as the economy proceeds on an innovative path. Consumers do not know the temporal structure of their preference system precisely, but know that its ordering can change together with the state of the economy. Therefore, the greater the degree of variability they anticipate in their preference system is, the larger the set of future options they will want to leave available will be, and hence the greater the amount of their income they want to keep in liquid form will be.

We assume in particular that, in each period, a consumers' utility function, whose arguments are the various commodities available on the market, is fully specified. Before determining the composition of their demand, however, the consumers fix the proportion of income that they want to devote to current consumption and the proportion they want to keep in liquid form. As we have already seen, the flexibility parameter σ, reflecting the consumers' long-term expectations as to the degree of variability of their preference system, determines P^* as the fraction of the money income available in each period actually devoted to the purchase of final output in the same period (the smaller σ, of course, the smaller such a fraction, and vice versa).

Once that has been decided, and given the prices $p_R(t+1)$ and $p_I(t+1)$ which are set by the producers and hence are parameters for the consumers, the composition of final output is determined as the solution of

$$\max \; U^* = R^{*\gamma^*} I^{*\delta^*}$$

which is the current-period consumers' utility function, under the budget constraint

$$P^*(t+1) = p_R(t+1)R^*(t+1) + p_I(t+1)I^*(t+1)$$

$$R^*(t+1) \geqslant 0 \qquad\qquad I^*(t+1) \geqslant 0.$$

We have already pointed out that on an innovative path the consumers' preference system undergoes a change as the result of a greater diffusion of products and/or services which they get to know better and better. A first effect of the greater acquaintance with new forms of consumption – which the consumers also help to define – is an increase of the degree of preference for the output of the innovative processes, that is of the parameter δ^* of the current-period utility function. We can refer to the cumulated output sold up to the period considered as an indication of the degree of diffusion of the new products, and write

$$\delta^*(t+1) = \delta^* \left(\sum_{T=n+1}^{t} \min \{I(T), I^*(T)\} \right)$$

$$\delta^*(n+1) = \overline{\delta}^*, \; \Delta\delta^*(t+1) > 0$$

assuming that

$$\sum_{T=n+1}^{t} \min \{I(T), I^*(T)\} > \sum_{T=n+1}^{t-1} \min \{I(T), I^*(T)\}$$

where $\overline{\delta}^*$, exogenously determined, is the degree of preference for the output of the innovative processes when it first appears on the market and the consumers do not know anything about it (under the hypothesis that an innovative choice has first been made at time 0).

From $\delta^*(t+1) > \delta^*(t)$, and given that $\delta(t+1) = e\delta^*(t+1) = \delta^*(t)$, it follows that $\delta^*(t+1) > \delta(t+1)$. If this effect is not more than compensated by a great enough difference between $P(t+1)$ and $P^*(t+1)$, the output of the innovative processes is always fully absorbed by the market, that is, $I^*(t+1) > I(t)$.

A greater acquaintance with the new products, however, might not only attract an increasing share of the fraction of income actually devoted to the purchase of final output, but might also bring about an increase of this very fraction. It might in other words have an effect on the long-term expectations of the consumers, by increasing the degree of confidence in the market and thus reducing the degree of variability they anticipate in their preference system (i.e. their search for flexibility and hence for liquid assets) represented by the parameter σ. We can thus make a reduction in the fraction of the available income actually subtracted from consumption a function of the diffusion of the output of the innovative processes, and write

$$\sigma(t+1) = \sigma \left(\sum_{T=n+1}^{t} \min \{I(T),I^*(T)\} \right) \qquad \Delta\sigma(t+1) > 0.$$

The Solution of the Model

The solution of the model in each period – and hence, given the links that join up the successive periods in a sequence, the pattern of evolution of the economy – is given by the values of the vectors x^u, y^u, x^c, y^c and of the scalars x_0, y_0, which, given the values of the parameters and of the exogenously determined variables of the model, satisfy the existing constraints.

In particular, the values actually taken will depend on whether the financial constraint or the human resource constraint is the more stringent.

Under the Financial Constraint   Let us first consider the case in which the financial constraint prevails. Given the Wages Fund and the state of the expectations, decisions as to current production and investment determine the size and the structure of the demand for labour defined by the vector $L^D_{s+1}(t+1)$.

The decisions as to current production, given the routine and the innovative processes already in the utilization phase inherited from the past, determine x^u, y^u and u^u, v^u, according to the scrapping rule adopted.

In particular we assume that the scrapping of the processes already in the utilization phase depends on the age structure of the inherited productive capacity and on the gap between the latter and the amount of final output decided for the current period; that is

$$u^u_k(t+1) = F_{kx}(x^u_{m+1}(t), x^u_{m+2}(t),\ldots, x^u_{m+M-1}(t), x^c_m(t); D_R(t+1))$$
$$k = m+1, m+2, \ldots, m+M$$

where

$$D_R(t+1) = c'(\boxed{J}^u_x x^u(t) + E^u_x x^c_m(t)) - R_e(t+1)$$

for the routine processes, and

$$v_k^u(t+1) = F_{ky}(y_{n+1}^u(t), y_{n+2}^u(t), \ldots, y_{n+N-1}^u(t), \ y_n^c(t); D_1(t+1))$$
$$k = n+1, n+2, \ldots, n+N$$

where

$$D_1(t+1) = d'(t+1)(\boxed{J}_{y}^{u}y^u(t) + E_{y}^{u}y_n^c(t)) - I_e(t+1)$$

for the innovative processes ($\boxed{J}_x^u$, $\boxed{J}_y^u$ being matrices of the type

$$\begin{bmatrix} 0 & 0 \\ \boxed{I} & 0 \end{bmatrix}$$

of dimension $M \times M$ and $N \times N$ respectively, and E_x^u, E_y^u column vectors of the type

$$\begin{bmatrix} 1 \\ 0 \\ 0 \end{bmatrix}$$

of dimension M and N respectively).

The functions F are such that the older processes are scrapped first. They reflect a flexibility criterion that focuses on expected final output as an index of a less stringent expected financial constraint, and hence of a greater range of future options.

The processes of the two types in the utilization phase, to which labour keeps being applied in period $t+1$, will then be

$$x^u(t+1) = \boxed{J}_x^u x^u(t) + E_x^u x_m^c(t) - u^u(t+1)$$

$$y^u(t+1) = \boxed{J}_y^u y^u(t) + E_y^u y_n^c(t) - v^u(t+1)$$

where the products $\boxed{J}_x^u x^u(t)$ and $\boxed{J}_y^u y^u(t)$ reflect the fact that the processes in the $(m+M)$th, $(n+N)$th period of the utilization phase in period t disappear in the period $t+1$, while the products $E_x^u x_n^c$ and $E_y^u y_n^c$ register the processes that pass from the construction to the utilization phase in the period $t+1$.

Decisions on investment are constrained by the remaining financial resources which determine the scrapping of the processes in the construction phase as well as the current rates of starts of new processes. The rates of starts are given by

$$x_0(t+1) = G_{0x}(x_1^c(t), \ldots, x_{m-1}^c(t), y_1^c(t), \ldots, y_{n-1}^c(t), x_0(t), y_0(t), \Gamma(t+1))$$

$$y_0(t+1) = G_{0y}(x_1^c(t), \ldots, x_{m-1}^c(t), y_1^c(t), \ldots, y_{n-1}^c(t), x_0(t), y_0(t), \Gamma(t+1)).$$

The processes scrapped are given by

$$u_k^c(t+1) = G_{kx}(x_1^c(t),\ldots,x_{m-1}^c(t),y_1^c(t),\ldots,y_{n-1}^c(t),\ x_0(t),y_0(t),\Gamma(t+1))$$
$$k = 1,2,\ldots,m$$

$$v_k^c(t+1) = G_{ky}(x_1^c(t),\ldots,x_{m-1}^c(t),y_1^c(t),\ldots,y_{n-1}^c(t),\ x_0(t),y_0(t),\Gamma(t+1))$$
$$k = 1,2,\ldots,n$$

where

$$\Gamma(t+1) = W(t+1) - w'(t+1)(\boxed{A}^u x^u(t+1) + \boxed{B}^u(t+1)y^u(t+1)).$$

The scrapping functions G are such that the younger processes are scrapped first: they are in fact the less irreversible processes, that is the processes that have so far swallowed the least resources, and are also further from the moment at which they will come up to final output. Then we obtain $x_0(t+1),y_0(t+1)$, which are the first to be cut, and

$$x^c(t+1) = \boxed{J}_x^c x^c(t) + E_x^c x_0(t) - u^c(t+1)$$

$$y^c(t+1) = \boxed{J}_y^c y^c(t) + E_y^c y_0(t) - v^c(t+1)$$

where the products $\boxed{J}_x^c x^c(t)$ and $\boxed{J}_y^c y^c(t)$ reflect the fact that the processes in the mth, nth period of the construction phase in period t pass to the utilization phase in period $t+1$, while the products $E_x^c x_0(t)$ and $E_y^c y_0(t)$ register the new-born processes in the same period.

To sum up: the processes of production are scrapped, if necessary, in an order reflecting a flexibility criterion that focuses on expected final output (both its amount and its nearness in time) as an index of a less stringent expected financial constraint. In this perspective, the distance from the moment the final output will first be obtained becomes relevant for the processes still in the construction phase, and the length of time over which it will still be obtained for the processes already in the utilization phase. Thus, when there is a reduction of resources, the first to be cut will be the rate of starts of new processes, then the processes in the construction phase (first the younger, then the older), and last the processes in the utilization phase (first the older, then the younger).

Under the Human Resources Constraint   When the structure of the demand for labour that results from the solution given by the values of

$x^u, y^u, x^c, y^c, x_0, y_0$ under the existing financial constraint does not coincide with the structure of the labour supply, as happens when an innovative choice is made, a human resource constraint appears.

However, we assume that the same equations as determined x^u and y^u under the financial constraint continue to do so. As a matter of fact in a first phase – that is the construction phase of the innovative processes – current production concerns only the routine output, which is obtained by means of processes to which the existing labour force is already completely adjusted. After that, that is when the output of the innovative processes starts appearing on the market, it is unrealistic to consider a human constraint so stringent as not to allow current production to be carried on.

Only the investment decisions, therefore, actually reflect the appearance of a human resource constraint – in the scrapping equations which become

$$u_k^c(t+1) = H_{kx}(x_1^c(t),...,x_{m-1}^c(t),y_1^c(t),...,y_{n-1}^c(t),x_0(t),y_0(t);\hat{L}_{s+1}^D(t+1))$$
$$k = 1,2,...,m$$

$$v_k^c(t+1) = H_{ky}(x_1^c(t),...,x_{m-1}^c(t),y_1^c(t),...,y_{n-1}^c(t),x_0(t),y_0(t);\hat{L}_{s+1}^D(t+1))$$
$$k = 1,2,...,n$$

$$x_0(t+1) = H_{ox}(x_1^c(t),...,x_{m-1}^c(t),y_1^c(t),...,y_{n-1}^c(t),x_0(t),y_0(t);\hat{L}_{s+1}^D(t+1))$$

$$y_0(t+1) = H_{oy}(x_1^c(t),...,x_{m-1}^c(t),y_1^c(t),...,y_{n-1}^c(t),x_0(t),y_0(t);\hat{L}_{s+1}^D(t+1))$$

where $\hat{L}_{s+1}^D$ is calculated taking into account that the structure of the demand for labour in each period varies not only according to the mix between routine and innovative processes, but also according to the age structure of the innovative processes themselves. The scrapping rules, on the other hand, are assumed to be the same as in the case in which the financial resource constraint prevails.

The appearance of a human resource constraint renders impossible a full use of the available financial resources, so the Wages Fund becomes

$$W(t+1) = w'(t+1)\hat{L}_{s+1}^D(t+1).$$

This is so if money wages are kept constant in time. Conversely, if the wages are allowed to increase in response to the appearing labour scarcities, all financial resources devoted to production can be actually employed. However, these implications for economic policy will be discussed at length in the following chapter.

5

Patterns of Evolution of the Economy: Issues and Policies

ECONOMIC POLICY IMPLICATIONS

Economic Policy as Stabilization Policy

The model outlined in chapter 4 provides a suitable tool for the analysis of phenomena that occur sequentially. In fact this model makes it possible to trace the patterns of evolution of the economy under alternative hypotheses as to the value assigned to the exogenously determined variables and the parameters, and to throw light on relevant aspects of this evolution. This in turn provides important insights into the working of the economy, thus making it possible to single out its relevant moments and their connections and to understand what policy interventions are required.

Before using the model as a tool for performing numerical experiments that make it possible to follow the sequence of the events through time, it must however be stressed that the model itself is the expression of a change of perspective in the analysis of problems of economic change, and that it provides an altogether different approach both to analysis and to policy implications.

Traditional policy approaches consider any change that can be interpreted as a break in a pre-existing situation as something that can be taken care of in purely quantitative terms. They are in fact inspired by models that, for the most part, have been built to obtain (or to maintain) the equilibrium between the demand generated by the economy and its supply capacity through a mere quantitative adjustment. As already pointed out in chapter 2, in these models a

79

given productive capacity, which is the expression of a given 'technique', can be made to work at different levels of intensity, or augmented/reduced, in order to meet an increase or decrease in demand. Whether we consider a given capacity or changes in its dimensions, the problem is to increase or to decrease the quantities produced in order to fill the gap between supply and demand, which appears either at the level of final demand or at some intermediate stage of production.

In this context, every break in a pre-existing equilibrium situation is interpreted as an instability problem, and accordingly economic policy is conceived as stabilization policy. The underlying idea, in fact, is that there is an equilibrium position (point or path) to be reached or returned to, so the fluctuations in output and in employment that affect the economy are considered as disturbances that have an essentially *reversible* character. This in turn implies the interpretation of these disturbances as purely quantitative phenomena, not only at an aggregate level, but even when they imply structural modifications in the particular sense of changes in the composition of the aggregate.

This character is common both to policies usually defined either as 'structural' or 'conjunctural', and also to Keynesian and monetarist policies, just to quote the better known examples.

The Keynesian Approach

The typical instability problem in Keynesian analysis is the instability of Harrod's model, which arises from a difference between the existing and the desired stock of capital because of a shortage of final demand. If the growth rate of investment is then decreased below the expected growth rate of output in order to eliminate this difference, the result is a less than proportional growth of output, which increases the gap between existing and desired capital stock (Hahn and Matthews, 1964). The state of excess capacity is thus self-reproducible and the shortage of demand is aggravated from one period to the next. It can be clearly seen that this chain reaction can only be stopped by injecting demand into the economy from outside, and this is what Keynesian policy essentially comes down to.

More recent developments of the analysis point to sectoral disequilibria, which become responsible for the simultaneous appearance of inflation and unemployment, as to the result of an instability that, however, is interpreted no longer as a cumulative mechanism leading the economy further and further from the equilibrium position, but as a disequilibrium that cannot be reabsorbed. The analysis still focuses on the quantitative aspects of this instability.

Tobin (1972) considers different markets characterized by different disequilibrium situations of opposite sign – and unemployment is then defined as a disequilibrium phenomenon. Money wages fail to adjust rapidly enough to clear all labour markets each day: excess supplies lead to unemployment and excess demands to unfilled vacancies. At any given moment markets vary widely in excess demand or supply, and the economy as a whole has both vacancies and unemployment side by side. The simultaneous occurrence of vacancies and unemployment is a measure of the heterogeneity or the dispersion of individual markets, and the amount of dispersion depends directly on exogenous shocks (changes in demand and/or in technology) that keep the market in perpetual disequilibrium. Given the level of aggregate demand, the larger the variance in excess demand or supply is between markets the greater the wage inflation will be.

The problem is clearly expressed as a mismatch between supply and demand in the various sectors, which calls for a redistribution of resources (labour) and demand.[1] The main point, in this context, is whether an equilibrium position to get back to exists (can be defined) or not.[2] If it exists, quantitative adjustment through a redistribution of resources between the various sectors should, in principle, lead the economy back to equilibrium; if this does not happen, it is either because the economy is continuously subjected to new random and exogenous shocks or because of price and/or wage rigidity which most of the time reflect non-competitive conditions.[3]

Of course, what lies behind all this is a horizontal image of the economy, made up of sectors that coexist, at the same time and at the same level, within which to bring about a redistribution of resources. Consumption and investment, in this context, are treated as sectors that have the same analytical status; and this is the main reason for the standard Keynesian approach being unable to take explicitly into account policy problems associated with the construction of a new productive capacity, that is with the phase during which the capacity itself acquires its distinctive character (see chapter 2 above).

[1] 'Workers will move from excess-supply markets to excess-demand markets and from low-wage to high-wage markets. Unless they overshoot, these movements are equilibrating. The theory therefore requires that new disequilibria are always arising' (Tobin, 1972, p. 9).

[2] The existence of an equilibrium position to which to refer, in fact, makes it possible to interpret any disturbance, whatever its nature, as a quantitative phenomenon.

[3] The most advanced expressions of the argument on price rigidity are the models with quantitative rationing of the Drèze (1975), Benassy (1975) and Malinvaud (1977) type. Price formation under monopolistic competition has been analysed by, among others, Benassy (1976), Grandmont and Laroque (1976), Negishi (1977) and Hahn (1978).

The Monetarist Approach

The difference between monetarists and Keynesians – as regards the problems that their underlying models allow one to focus on – is not a difference in their nature but a question of degree. In fact, one could go as far as to say that in this respect the monetarists make explicitly clear what is implicit in the Keynesian approach.

The most recent monetarist theses, which attribute the instability of the economy to monetary changes and, in particular, make the price level depend only on the money stock, refer to a rational expectation equilibrium – that is, a Walrasian equilibrium of a perfectly competitive economy in which there is no learning – which the real economy can be out of only for short periods (Hahn, 1982). The real values of the relevant variables of the economy can diverge from the values acquired in an equilibrium position such as this only if agents make mistakes, and economic policy can move the real variables from their equilibrium values only if it has a random component, that is, if the agents are taken completely by surprise. Any regularly implemented policy (whether monetary or not) in fact becomes part of the environment, can be read by the agents and reacted to in such a way as to cancel its effects.

It follows that the changes that affect the real variables can only be strictly reversible changes of a monetary origin, of a quantitative nature and of a random character, and that the policies accordingly proposed (namely, a fixed monetary growth rule to prevent the very occurrence of random monetary changes (Friedman, 1971)) are explicit quantitative policies.

The reasons for advocating a passive economic policy are on the other hand made crystal clear by Lucas (1975), according to whom an active policy would have the effect of 'retarding the movements of resources into the most desirable activities' (1975, p. 1140). Redistribution of resources in other words (that is, a purely quantitative adjustment) is seen as the essence of the problem whatever the kind of disturbance considered and whatever its origin; and this, as we have seen, is exactly what also lies behind the standard Keynesian approach and the policies inspired by it. Thus, the only difference between the two approaches comes down to whether active or passive policies are considered more suited to achieving this result.

'Out-of-Time' Economic Policies

Keynesian policies are policies for regulation of demand so as to allow it to match, as exactly as possible, potential output. Their essential aim is the correction of demand behaviour, and this is pursued through a mechanism (multiplier – accelerator) that focuses on the level of demand as the only relevant aspect. The most distinctive expression of this approach is the consideration of investment at the same level as consumption, as a starter of that quantitative adjustment process that is the multiplier.

In this representation no attention is really paid to the fact that 'production takes time'; but, as Hicks (1974a) pointed out, 'it is impossible to tell the multiplier story properly in terms of the flow relations between income and saving, to which Keynes (in the main) confines himself. The state of stocks, even the initial state of stocks, must be considered too' (p. 14). Thus when stocks are taken into account in the analysis, as is always the case when productive capacity (and a modification of it) is considered, it is no longer possible to ignore the fact that the process of change takes place in real time. Productive capacity must be created before it can be utilized, and this has important implications in the analysis of structural changes. This implies in particular that investment, when changes in this capacity are taken into account, cannot be considered at the same level as consumption. Therefore when Keynesian policies are implemented in situations in which the structure of productive capacity is not adjusted to the structure of demand, not only do they not get the expected results in terms of output and employment but they may cause inflation and a greater indebtedness. This is usually the moment when monetary policies are considered.

In the monetarist approach these policies should only aim at substituting financial assets for real assets in order to get an equilibrium asset structure. However, a restrictive monetary policy also amounts to a more stringent financial constraint, which implies the destruction of part of the existing productive capacity that leaves the producers with the corresponding debts. In this case the implication of the consideration that production takes time is that it is not possible to interrupt the construction or even the utilization of a given productive capacity without incurring a cost. Monetarists do not consider this aspect because they abstract completely from all kinds of stocks except money stocks, as Hahn (1982) pointed out.

In conclusion, monetarist as well as Keynesian policies are unsuitable for dealing with problems of economic change when these imply structural modifications, as is typically the case with technical change and innovation. The reason, as we have stressed, is represented by the purely quantitative, reversible character of the policies devised to get or to return to an equilibrium position that can be defined *a priori*.

ROUTINE CHOICE VERSUS INNOVATIVE CHOICE

A 'Viability' Problem

In a quantitative adjustment perspective of the problem of economic change, the decision as to the option to follow comes down to a choice between *better* and *worse*, where the best choice usually reflects some sort of maximization criterion.

Our outlook, as we have already pointed out, is altogether different. We consider the case for a qualitative change that implies *different* answers to different production (and consumption) problems, which can be clearly defined only through a process in time. The way in which the economy actually functions, embodied in the existing productive capacity, depends on the skills of human resources which reflect the characteristics of the environment. A modification of the environment as a source of technology is then the outcome of a learning process which changes the characteristics of human resources, and can only come about as the result of an innovative choice that triggers off a process of innovation interpreted as a sequential process of research and experimentation. In such a context, which is fundamentally uncertain, the idea of a choice between given alternatives whose outcome is more or less known is meaningless.

All we can do, and what it is relevant to do, is analyse the various patterns of evolution of the economy that correspond to various choices (in our case, a routine choice and an innovative choice), in order to be able to deal with the problem of the viability of an innovative choice, which is the only appropriate answer when a structural modification is considered. For the viability of such a choice we need: (1) not to be faced with a lack of financial resources such as to hamper the functioning of the economy, and (2) not to be faced with intolerable levels of indebtedness (that is, to the point that would make it no longer possible to go on substituting external financial resources for the internal ones) and of unemployment, when this would imply reductions in the amount of human resources involved in innovative processes of production, and thus affected by the learning process, such as to bring about in turn serious reductions in the intensity of this very process.

The setting of the analysis is a steady growth sustained by constant (and consistent with each other) growth rates of the exogenously determined magnitudes of the economy: namely, the inflow of 'money' ΔM, consumption out of profits Q, and the supply of the various types of labour inputs represented by the various elements of the labour availability vector L_s^s. The economy makes use of an already established and fully developed technology to satisfy a final demand that reflects a given consumer preference system. Everybody is perfectly satisfied with what is going on and thus there is no search for flexibility either on the part of the consumers or on the part of the producers ($\sigma, \varrho = 1$).

At a given moment $T = 0$, saturation of demand for the existing commodities and/or the perception of the possible exploitation of new (scientific) knowledge is assumed to reduce the confidence that the consumers and/or the producers have in the existing state of affairs, thus bringing about a modification of their long-term expectations. There will therefore be a break in the equilibrium sequence and the beginning of a search for flexibility which will result in a decrease in the value of σ and/or ϱ (see chapter 3 above). This, in turn, will bring about directly or indirectly (through a reduction of the Wages Fund) a reduction in final demand that will result in a flow disequilibrium in the current period.

This disequilibrium can either be (wrongly) considered by the producers as the result of a mistake in formulating short-term expectations, or (rightly) regarded as the signal of a structural change, that is, as the result of a modification of long-term expectations.

In the first case the producers give a quantitative interpretation to the appearance of an excess supply, and consequently their reaction consists in a simple revision of the final production and investment targets of processes of production still carried on according to the established technology. The economy continues to follow a *routine* path. In the second case the producers give a qualitative interpretation to what is going on, and their answer then consists in a revision of the decisions on *how* to invest, by setting off on an *innovative* path.

The analysis of the patterns of evolution of the economy in the two cases will make it possible to throw light (1) on the implications of the consideration that 'production takes time', when quantitative economic policies are the answer to a problem that calls for a qualitative change, and (2) on what is required for an innovative choice to be viable, when conversely the latter is considered.

This analysis, as we have already said, will be developed by means of simulations that will make it possible to see what would happen, conditional on alternative assumptions in terms of the exogenous variables and the parameters of the model. They will be non-stochastic simulations in that the exogenous variables and the parameters will not be set at their expected or most likely values by using random samples, but at values in various scenarios whose consequences we would like to evaluate. In fact we are not interested in the construction of confidence intervals as would be the case if we were looking after the properties or the performance of the model with respect to some equilibrium point or path, or the robustness of the economy to stochastic shocks.

Our real aim is to explore the articulation of the sequence in time that represents the evolution of the economy under alternative hypotheses, in order to bring to light the relevant moments and connections of this sequence. In this perspective we have drawn various scenarios corresponding to different hypotheses as to the nature and the strength of the existing constraints, as to the behaviour of economic agents in their search for flexibility and as to the economic policy followed.

The Routine Choice

The evolution of the economy on a routine path – given the way in which, in a quantitative adjustment perspective, the producers are expected to react to the appearance of a flow disequilibrium – depends mainly on the particular policy that the Government chooses to follow. This policy, in the model proposed, is expressed by the rule according to which the inflow of money ΔM is exogenously determined. Changes in Q (which represents consumption out of profits) contribute to the determination of the growth rate of the economy as well as ΔM, but in a different way. ΔM, in fact, goes to increase the Wages Fund, while Q directly affects final current demand while reducing the Wages Fund.[4]

In fact, on a routine path the labour force is already perfectly adjusted to the prevailing technology, and thus there is no labour constraint. The only existing constraint is therefore represented by the available financial resources that determine the levels of activity of the economy and whose amount and destination are affected mainly by the values taken by ΔM (and Q).

Two scenarios have been drawn in consequence. In the first one we assume that the fall in the growth rate brought about by the initial disequilibrium is accepted, and that the growth rate of ΔM (and Q) is adjusted to the lower level resulting from the initial fall in final demand and hence forward adjusted period after period to the actual growth rate of the economy whatever this happens to be (a kind of feedback rule). The aim, in this case, is to stabilize the economy, in the sense of stopping a cumulative downward mechanism. In the second scenario ΔM (and Q) are kept growing at the original rate

---

[4] Exogenous changes in Q, on the other hand, could be made to represent changes in public consumption, which could then be given in the model the same 'take-out' status as consumption out of profits.

whatever the actual growth rate of the economy (constant rule); thus trying to return to the potential rate at which the economy was originally faring.

Stabilizing the Economy The numerical experiments performed show that in the first scenario – once the values of ϱ and/or σ are reduced in period $T=0$ and as a consequence a flow disequilibrium appears in the same period – adjustment of the growth rate of ΔM (and Q) to the present lower level makes it possible to re-establish the supply-demand equilibrium from the next period onwards. The growth of the economy is thus immediately stabilized, unless of course there are further reductions in ϱ and/or σ^5 (see figure 1).

However, a lower growth rate implies that some processes of production in the utilization phase must be scrapped, and this will go on as long as the inherited productive capacity is not fully adapted to the expectations of final demand, that is, as long as the economy remains in a state of stock disequilibrium (figure 3). During this period, the funds that were invested in those processes are lost as the processes themselves cease to exist; but the producers are left with the corresponding debts, so this stock disequilibrium implies in turn an increase in the degree of indebtedness. More generally, the reduction of the growth rate of ΔM to the lower growth rate of the economy cannot be immediately matched by a similar reduction in the flow of repayments which, at least for a while, remains at the original level, continuing to run above the growth rate of the economy; as a consequence the growth rate of the external resources becomes greater than that of the internal resources.

Finally, unemployment of all types of labour increases with an unchanged technology, as the growth rate of labour demand that reflects the new lower growth of the economy becomes smaller than the unchanged growth rate of labour supply that results from demographic and educational factors (see figure 2). The excess supply of labour, in an economy in which the equilibrium on the final-output market is immediately re-established, could be labelled as 'Keynesian unemployment'. In fact, in the model it depends on a shortage of P, which is the current-period final demand expected by the producers, that is, the Keynesian 'effective demand'. However, this shortage of demand depends on a structural change (a modification of long-term expectations), so the unemployment that results from a growth rate

---

^{5} This would be the case if the current-period flow disequilibrium also had an effect on long-term expectations controlling investment, as happens in case of a Harrodian instability.

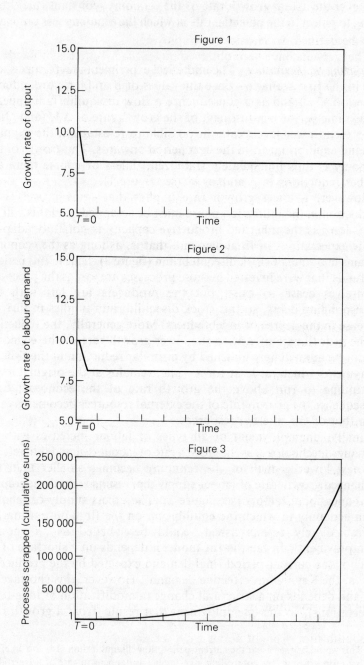

Routine path – first scenario

of productive capacity that, given the scrapping made at the beginning of each period, is smaller than the growth rate of labour supply, could also be defined as 'classical unemployment'.

These results have been obtained under the hypothesis of fixed real wages which, given the way in which prices are determined in the model and on the basis of the assumption made in the simulations that prices are perfectly rigid downwards ($K_R = 0$),[6] also implies fixed money wages. One might therefore be tempted to think that a decrease in real (and money) wages is a solution to the unemployment problem on a routine path. In fact, given the Wages Fund, lower real wages allow higher rates of starts of new processes and hence higher employment, provided of course that the decrease in the wages does not bring about a revision of producers' long-term expectations and a reduction of ϱ, as would be the case in a Keynesian perspective.

In any case, for the given pattern of evolution of final current demand, the creation of employment associated with the starting of a greater number of new processes of production will necessarily be followed, when the moment comes, by the destruction of employment resulting from a correspondingly greater number of processes scrapped. The net balance depends in part on the ratio of the labour inputs of the construction phase to those of the utilization phase. What is clear, in any case, is that the creation of employment through a wage reduction implies a greater scrapping of production processes and a greater indebtedness for the producers.

The excessive growth of the indebtedness is the real problem that the economy has to face on a routine path. Sooner or later in fact this will call for counter-measures which are represented in the model by a reduction in the growth rate of ΔM below that required for stabilizing the economy. Such a reduction means in fact a stronger financial constraint, a smaller Wages Fund, lower investments (that is, a reduction both in the rate of starts and the number of processes of production in the construction phase) and a fall in current production. The immediate consequence is the reappearance of a flow disequilibrium in the current period; the carrying of the resulting stock disequilibrium down the sequence will then cause the growth rate of the economy to keep falling in each successive period.

Getting Back to Potential Growth The attempt to bring the economy back to its potential growth rate is pursued in the model by keeping ΔM (and Q) growing at the same rate as before the appearance of the disequilibrium brought about by a reduction in ϱ and/or σ.

---

[6] In the scenario considered only excesses of supply occur, and therefore tendencies to a fall rather than a rise in prices.

The effects of this attempt to refuse a reduction in the growth rate of the economy have been considered in the first place with reference to the case portrayed in the model proposed in the previous chapter: that is, the case of a final output whose excess supply results in unwanted stocks that, just like the money stocks that result from an unsatisfied final demand, are automatically shifted to the following period(s). In this case the evolution of the economy is characterized by fluctuations in the growth rate which reflect excesses of supply and excesses of demand that alternate regularly. These fluctuations have a permanent (self-feeding) character, and their amplitude depends on the values of the parameters ϱ and σ which measure the producers' and the consumers' preference for liquidity (see figure 4). Below certain values (i.e. when the preference for liquidity is quite strong) the fluctuations acquire such an amplitude that after a certain number of periods all the processes of production in the construction phase must be scrapped because of the lack of financial resources (figure 4(b)).

In fact, the results of the numerical experiments performed suggest that two phases can be considered. In the first one the economy converges to the original growth rate through a damped oscillatory movement which reflects the fact that the supply–demand equilibrium cannot be immediately re-established – as in the first scenario drawn – and excess supply keeps alternating with excess demand as a result of the over- and underestimations made period after period by the producers in formulating their short-term expectations. In a second phase – when, after the construction phase, the oscillations in the rate of starts that have occurred during the first phase begin to produce their effects – the oscillatory movement becomes explosive, until the economy gets into an intolerable state of disarray: that is, until the levels of indebtedness and unemployment increase enormously (figures 5 and 6). However, convergence to the original growth rate in the first phase takes a very long time; this pushes the second phase out of sight, making the attempt to get back to the original growth rate of the economy appear as a viable option.

With a low preference for liquidity (values of ϱ and σ near unity) the fluctuations in the growth rate of the economy have a smaller amplitude. The fluctuations in the rate of starts of new processes of production are thus less pronounced, and a complete interruption in the construction of new productive capacity can be avoided. The existence of stocks of final output, which make it possible to dampen the effects of current disequilibria on the financial resources for investment, play an important role in this. In fact fewer resources are required for financing a current production that can be reduced in proportion to the existing stocks to satisfy a given demand, and more resources are correspondingly left available for the construction of new productive capacity.

When there are no stocks to take the strain of supply–demand inequalities – as happens when the final output is not storable, where the excess supply goes to waste and the excess demand is satisfied through imports – the path followed by the economy, whatever the value of ϱ and σ, is characterized by fluctuations that after a while become explosive, bringing about a shortage of financial resources which implies complete abandonment of the construction of new productive capacity.

However, the attempt to get back to potential growth by maintaining the growth of ΔM (and Q) at the original rate gets into difficulty not so much because it runs immediately into an absolute shortage of productive capacity, but because it induces an increasing variance in the *age structure* of the processes of production, which becomes a source of instability in the sense that it brings about fluctuations in the productive capacity that become more and more pronounced as time goes by.

In conclusion, the results obtained in both the cases considered (acceptance of a lower growth rate or not) show the substantial failure of policies conceived as purely quantitative interventions, at least when the structural change required, as revealed by the modifications in the agents' liquidity preference, is a profound one.

The Innovative Choice

On the other hand, the appearance of a flow disequilibrium in a given period can be correctly read as the symptom of a dissatisfaction with the existing situation on the part of the consumers (who might have reduced their demand because they want something *different*, even if not yet fully specified, and not just less of the existing commodities) and/or on the part of the producers (who might be in search of an altogether different, even if not yet well defined, approach to production), implying a modification of long-term expectations and hence of the flexibility parameters σ and/or ϱ. The way would then be open for the search for a qualitative answer (and not a simple quantitative adjustment) to a qualitative problem: that is, to the consideration of an innovative choice.

Thus an initial modification in long-term expectations that results in a fall in final demand in a given period, if correctly interpreted, cannot have an influence only on how *much* to invest in the following periods through a modification of the financial constraint brought about by a revision downwards of short-term expectations (which involves a reduction in current production and hence lower proceeds,

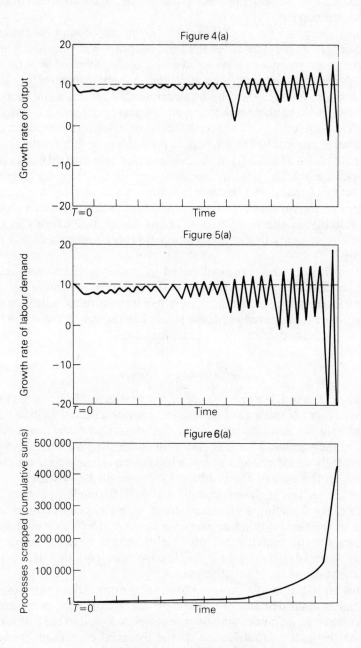

Routine path – second scenario
(weak search for flexibility; ϱ, $\sigma = 0{,}99$)

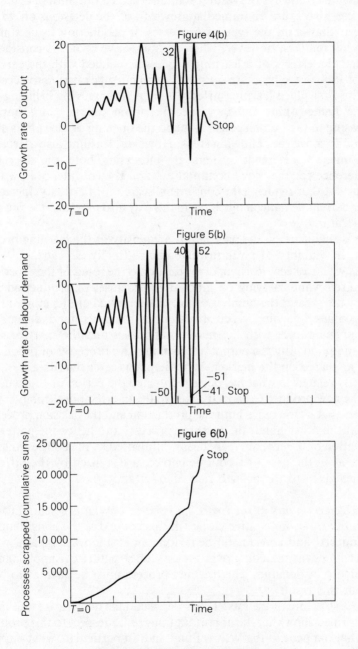

Routine path – second scenario
(strong search for flexibility; ϱ, σ = 0,95)

and which is always the result of an unexpected fall in final demand), but can also cause an immediate revision of the decisions on *how* to invest, that is on the type of processes of production to be started.

What matters, however, when an innovative choice is considered, is that the effects of a learning process associated with the carrying on of innovative processes of production must be taken into account. In the first place learning affects human resources by bringing into being higher skills and new specific labour inputs which imply a releasing of the human constraint and the opening of new paths along which to move the following steps. However, learning also affects the consumers' preference system, by increasing both the degree of preference for the 'new' commodities and the confidence in what is going on, thus reducing the consumers' search for flexibility, increasing their current demand and helping in this way also to release the financial constraint.

It is therefore on the pace and the intensity of the learning process that the viability of an innovative choice finally depends.

As we have seen, learning in production is the result of the experience acquired while carrying on innovative processes of production and, thus, the greater the number of processes carried on the more intense it becomes. Learning in consumption, on the other hand, depends on the acquaintance with the new commodities and the new forms of consumption after the output of the innovative processes of production has appeared on the market. What happens after that date is in turn strictly related to what has occurred during the period of construction of the new productive capacity, due to the intertemporal complementarities that characterize both the production and the decision processes.

On the other hand, the learning process, and hence the pattern of evolution of the economy, are deeply influenced by the policies followed that, as in the case of a routine choice, are stylized by the different values given to the growth rate of ΔM (and Q).

The Determinants of the Learning Process   What is essential for the viability of an innovative choice in this context is the consistency of producers' and consumers' behaviour on an innovative path, which determines the matching over time of the structure of supply and the structure of demand. The learning process must be such as to bring about this result.

The structure of the model considered and the numerical experiments performed show that the important thing, with respect to this problem, is what happens to the Wages Fund and, in particular, what happens to the rate of starts of new processes of production which, on an innovative path, determines not only what the level of activity of the economy and employment will be, but also how learning in both production and consumption will take place. A reduction in the Wages Fund in a given period, in fact, means a more stringent financial constraint, which, with fixed wage rates, involves a reduction in the rate of starts of new processes and, if the decrease in the Wages Fund is big enough, also the scrapping of some processes still in the phase of construction or already in the phase of utilization. This will result in the immediate slowing down of the process of learning, but also in a modification of the age structure of the productive capacity of the economy, whose effects will be felt over the future.

In the model, the rate of starts of new processes of production is constrained either by the financial resources available or by human resources, that is, by the particular labour requirements for carrying out an innovative choice.

Consider first the latter case, that is the case in which the human constraint prevails. Higher requirements for all types of labour in the phase of construction, on the one hand, imply a greater absorption of resources per unit process and thus a reduction in the number of processes of production that can be started with given financial resources. A change in the structure of the labour demand, on the other hand, even if it does not imply higher requirements for some or all of the various types of labour, causes the appearance of a labour constraint that can prevail over the financial one and thus reduce the rate of starts. This, as we have just seen, will immediately affect the learning process.

On the other hand, a lower rate of starts of new processes of production as a result of a change in the level or in the structure of the labour requirements when an innovative choice is made will at the beginning of the utilization of the new productive capacity always bring about a fall in final production, which implies a reduction in employment compared with what would have happened had the labour constraint not appeared. This effect on employment is similar to Ricardo's 'machinery effect' already mentioned. However, it does not necessarily depend on an increase in unit labour costs in the construction phase, as in Hicks's model (1973), but more generally on any change in the structure of labour demand such as to set a constraint on the rate of starts of new processes of production (see figures 8, 11(*a*), 11(*b*), 14, 17(*a*) and 17(*b*) later in this chapter).

The above-mentioned effects of an initial reduction in the rate of starts of new processes of production on an innovative path are likely to persist as long as there is a mismatch between the structure of supply and the structure of the demand for labour; they are gradually matched, however, by the process of learning in production associated with the carrying on of more and more innovative processes as time passes. The upgrading and the greater articulation of human resources as the process of innovation pushes on, in fact, implies a release in the human constraint – that is, in the model, a reduction in the labour input requirements together with a better matching of the structure of supply to the structure of demand, both of which allow a gradual increase in the rate of starts of new processes.

On the other hand a reduction in the Wages Fund – whether as a result of the appearance of a human constraint or not – also implies something else: less learning in consumption. A smaller Wages Fund means in fact lower incomes and a lower final demand, and this can lead to a situation in which the value of aggregate demand $P*$ is lower than the value of current supply P. If this difference is big enough it might more than compensate for the effect given by $\delta*(t+1) > \delta(t+1)$ and bring about a shortage of demand for the output of the innovative processes, which might not absorb all its current production, that is $I* < I$. Learning in consumption, which depends on the diffusion of the new output, would thus be reduced in turn.

A reduction in the Wages Fund is generally the result of a financial constraint – such as that brought about by a decrease in the value of the parameter ϱ which reflects an increase in the preference for liquidity on the part of the producers – or, as we have just seen, the outcome of the appearance of a human constraint, which sets a limit on the number of processes of production that can be carried on and hence, with fixed wage rates, also on the amount of financial resources originally devoted to production that can actually be employed. However, once the output of the innovative processes of production has started to flow onto the market, it can also come about as a result of the learning process itself. In fact, when the producers' behaviour is not consistent with that of consumers, this results in a flow disequilibrium in each given period (with the piling up of unwanted stocks of final output and corresponding shifts of demand, concerning the routine as well as the new output) which implies smaller proceeds in the same period and thus causes a reduction in the Wages Fund in the following period(s).

Thus learning in consumption brings about an increase in the value of the parameter $\delta*$, which expresses the consumers' degree of preference for the new output, and which in the model is perceived by the producers with a delay $\delta(t+1) = \delta*(t)$. In this case of a one-period delay in the adjustment of δ to $\delta*$ a flow disequilibrium cannot be avoided. This results in unwanted stocks of routine output and in shifts of demand for the 'new' output. The ensuing reduction in the proceeds of the producers and hence in the Wages Fund in the following period means a more stringent financial constraint, which involves a reduction in the rate of starts and perhaps also some scrapping of production processes, which implies in turn a greater indebtedness. On the other hand, this is only partially compensated by the expansionary effect on consumption due to the shifts in demand from period to period.

More generally the same thing happens in all the cases in which learning in consumption is such as to prevent the composition of final demand from evolving in accordance with the structure of productive capacity as determined by the evolution of the rate of starts and the scrapping undertaken. When this happens, in fact, a shortage of productive capacity cannot be matched immediately by a price increase, because this will only lead to a price change in the following period (bearing in mind that prices, in the model, are kept unchanged throughout each given period), and this necessarily implies smaller proceeds and a smaller Wages Fund in the following period.[7]

The fluctuations in the Wages Fund, which have been shown to be the main determinant in causing the learning process to advance by fits and starts, thus also appear to be the most important obstacle to the viability of an innovative choice.

The Evolution of the Economy on an Innovative Path   In the case of an innovative choice two scenarios can be drawn, in relation to different hypotheses as to the policy followed and under alternative assumptions as to whether the financial or the human constraint prevails.

In the first scenario the growth rate of ΔM (and Q), which stylizes the policy intervention, is made to follow the actual growth rate of the economy without delay. The numerical experiments show that when the financial constraint prevails this growth rate, after the initial fall, is immediately stabilized, as on a routine path. The consequences for indebtedness and for unemployment are likewise the same, that is, they continue to grow from one period to the next.

However, this is the case only until the end of the preliminary phase. After that, when the output of the innovative processes of production starts flowing on to the market and the effects of the learning process are increasingly felt, the growth rate of the economy can be augmented until it returns to its initial level, but this is accompanied by fluctuations in the demand for labour and in the scrapping of processes of

---

[7] Excess capacity cannot appear in the model because, at the beginning of each period, the required scrapping is always undertaken according to the expected final demand.

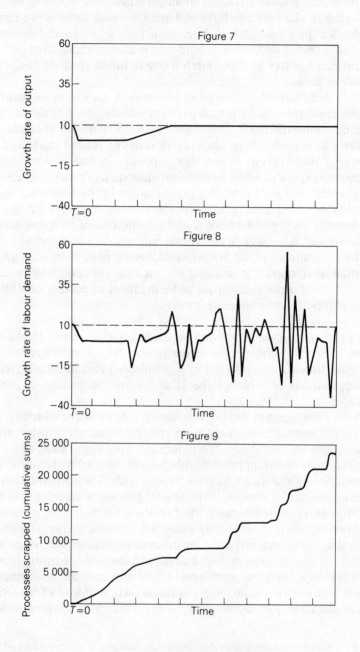

Innovative path – first scenario
(financial constraint)

production (due to fluctuations in the availability of internal financial resources) which are so strong as to cast doubts on the viability of the innovative path (see figures 7, 8, 9). This viability, in contrast, is certainly assured if ϱ and σ, thanks to the money stocks involuntarily accumulated in previous periods, can take values greater than unity, thus allowing an over-functioning of the economy (which can be made to grow still faster than it was originally doing) and hence a gradual reabsorption of unemployment. In this case the growth rate of ΔM (and Q) can be reduced below the actual growth rate of the economy (as the internal resources gradually become available to replace the external ones), thus also making possible a reduction of the indebtedness.

When there is a human constraint at work, on the other hand, not all the financial resources available for employment in production can effectively be used. The resulting fall in final demand triggers off a succession of excesses in supply and demand that reflects the way in which short-term expectations are revised in the model, and that causes fluctuations in the growth rate of the economy that, although damped, last all through the preliminary phase and even beyond – that is, up to the moment when learning in production causes the labour constraint to disappear (see figures 10, 11, 12). These fluctuations affect the learning process, but not so as to hamper the viability of the innovative path. However, they can be eliminated if, when a human constraint stronger than the existing financial constraint appears, money wages are increased in such a way as to permit the absorption of all the financial resources originally devoted to the Wages Fund, thus increasing the demand for current production and allowing it to match the supply capacity inherited from the past, period after period.[8]

The problem dealt with by wage increases is one of the relation between the financial constraint and the human constraint. The appearance of a human constraint is typically associated with an innovative choice: thus the stronger the producers' determination to set off on an innovative path (and hence the smaller their preference for liquidity, i.e. the nearer to unity the value of ϱ) the greater the constraint itself. With fixed wage rates, as is the case in the model, this can imply a stronger financial constraint than what would otherwise have prevailed, and hence the accumulation of greater money stocks than the producers are willing to hold. An increase in wages, in this context, appears as a way to reconcile the 'real' choice of the

---

[8] The increase in money wages must be such as also to bring about an increase in real wages, that is, in order to produce the above effect, it must be greater than the increase in prices.

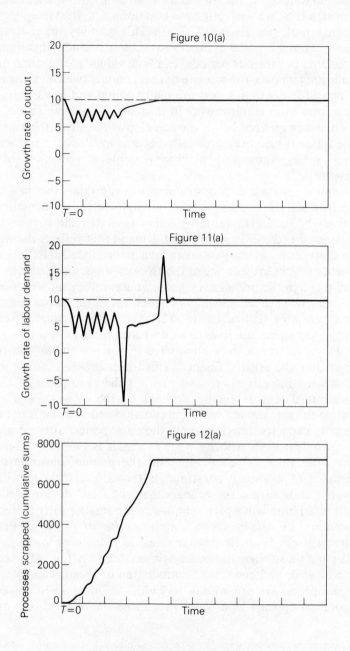

Innovative path – first scenario
(weak human resource constraint)

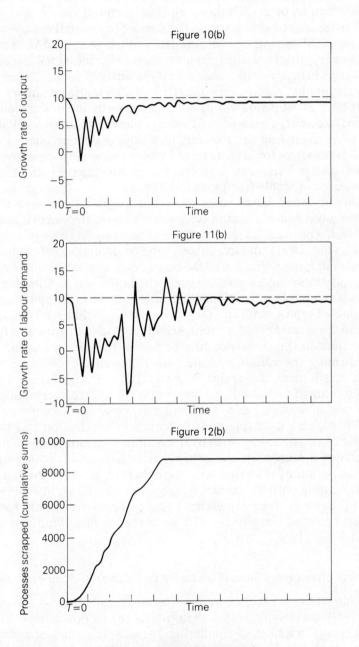

Figure 10(b)

Figure 11(b)

Figure 12(b)

Innovative path – first scenario
(strong human resource constraint)

producers – that is the choice as to the number of the processes of production to be carried on – with their financial choice.

The second scenario portrays the attempt to return the economy to its potential growth by keeping the growth rate of ΔM (and Q) unchanged, that is, at the level it had before the initial fall due to the reduction in the value of ϱ and/or σ. If the financial constraint prevails, the attempt has exactly the same effect as on a routine path. We shall thus have a first period during which the growth rate of the economy seems to converge to its original value, followed by strong oscillations that can either lead the economy to a deadlock on account of total lack of resources for investment or involve a scrapping so pronounced as to result in intolerable indebtedness. In this case the path followed is clearly not viable (see figures 13, 14, 15).

On the other hand, when the reduction in ϱ and/or σ is less pronounced and the human constraint prevails, the fluctuations are not so accentuated and the innovative choice may be viable (see figures 16(*a*), 17(*a*), 18(*a*)). In fact, although the predominance of the human constraint implies that not all the resources available for employment in production can be actually used, the initial fall in final demand thus brought about is at least partially compensated by the effort made by keeping ΔM (and Q) growing at an unchanged rate. Later on, in the second phase, the economy is even able to grow at a faster rate than in the corresponding period in the first scenario, thus confirming the viability of the path followed.

If the human constraint is particularly strong, however, the fluctuations in final output may not only lead to fluctuations in the rate of starts of new processes but, given the scrapping rules followed, also a scrapping of processes of production still in the phase of construction, such as to result in a serious increase in the degree of indebtedness and, what is more relevant, in a lengthening of the period of time that will elapse before the 'new' output will start flowing onto the market (see figures 16(*b*), 17(*b*), 18(*b*)). In this case the rate at which the external financial resources would have to be kept growing actually casts doubts on the viability of the innovative path.

An Economic Policy Conducive to Change and Innovation

An innovative choice, as we have seen, depends first of all on the interpretation that is given to the appearance of a flow disequilibrium that results in a stock disequilibrium (namely, in a stock of processes of production no longer consistent with the existing expectations). When producers are convinced that the break in the sequence requires

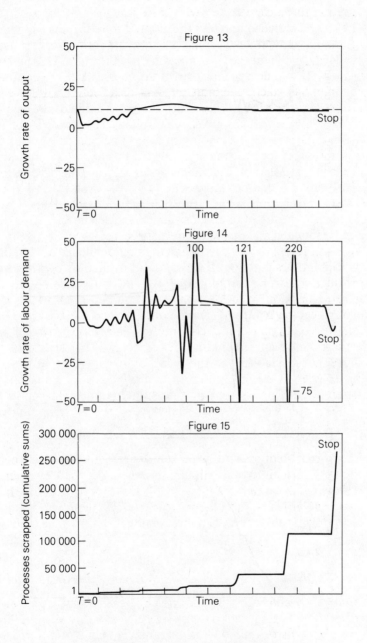

Figure 13

Figure 14

Figure 15

Innovative path – second scenario
(financial constraint)

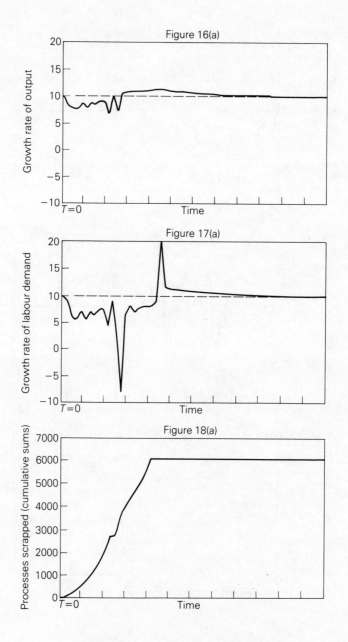

Innovative path – second scenario
(weak human resource constraint)

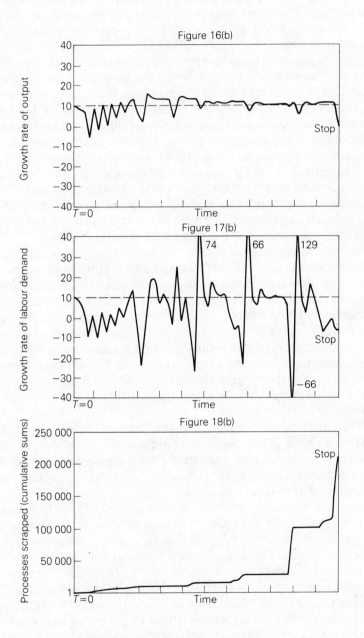

Innovative path – second scenario
(strong human resource constraint)

abandonment of the existing model, the search for flexibility that characterizes their behaviour in the uncertain context of a changing environment naturally leads to an innovative choice as to the required qualitative change.

In this perspective economic policy is no longer a policy for making a choice, but one aimed at rendering the innovative choice 'viable' according to given criteria. In the economy that we have portrayed the viability of a choice depends essentially on the availability of the financial resources required in order to avoid a possible stop in the functioning of the economy that can come about, precisely because of a lack of resources, in the course of the process of innovation; and this, as we have already pointed out, refers not so much to the *amount* as to the *moment* when these resources are required. Furthermore, a viable choice cannot be associated with intolerable levels of indebtedness and/or unemployment, which are the two evils that the economy is most likely to face when it undergoes a structural change.

Viability in these terms clearly depends on the consistency of the evolution of the structures of demand and supply on an innovative path. This does not mean, however, a quantitative reallocation of resources and demand between different sectors so as to ensure the equilibrium in a traditional sense. What is implied, in fact, is keeping in balance learning in production, which determines the changes in the structure of productive capacity, and learning in consumption, which determines the changes in the structure of final demand. Learning is the essence of the process of innovation. Policy interventions thus, in the end, must essentially affect the intensity and the pace of the learning process.

When defined as such – and when learning is rightly seen as concerning both production and consumption – an economic policy conducive to change and innovation clearly appears as an integrated set of active interventions which must take into account the interaction over time of supply and demand.

The first point to stress is that attention must not be so much concentrated on the quantitative content of these interventions (e.g. on the level of the money injection that, in the model, expresses the policy followed). The most important things for the viability of a process of innovation conceived as a sequential process of transformation of the productive structures of the economy are the implications of the consideration that production takes time. Accordingly, the most important aspect of the policy intervention is its articulation through time, that is not so much *how much* as *when*.

This appears clearly in the case that is most favourable to the viability of an innovative choice, that is in the first scenario drawn. In this case, in fact, the growth rate of ΔM (and Q), which stylizes the policy intervention, must be first reduced and then increased, starting from the moment when the new output begins to flow on the market, so as to accompany the actual growth rate of the economy. What matters, however, is not so much the stabilization of the growth rate of the economy thus obtained as the effects on the Wages Fund, and hence on the rate of starts and on scrapping, aimed at avoiding perturbations in the age structure of productive capacity, which might be a source of fluctuations that could affect the viability of the innovative path. Later on – if (and when) an increase in the value of ϱ and σ above unity allows an over-functioning of the economy – a reduction in the growth rate of ΔM below the actual growth rate of the economy is required in order to make possible a reabsorption of the indebtedness and of unemployment.

In the same way changes in wages as an instrument of economic policy must be correctly shaped in time, as the results of the model clearly show. Wage increases may be useful to counteract the effects on the rate of starts and on scrapping of a reduction in the Wages Fund caused by the appearance of a human constraint in the initial periods of a process of innovation, when this type of constraint usually prevails, while later on, when the learning process had helped to release the human constraint and the level of activity is once again limited by the financial resources available, such a policy would only bring about lower investments and hence greater unemployment, so it might instead be convenient to reduce the wage rates, provided of course this does not also bring about a revision of the long-term expectations of the producers in the wrong direction.

Essentially, this change of perspective in economic policy reflects the difference in the analytic status of consumption and investment that comes to light when the time dimension of production is taken explicitly into account. In this perspective, as we have seen, it may be convenient to sustain current consumption not for its own sake but for the effects that this can have on the process of creation of the productive capacity of the economy and hence on the structure of that capacity.

What is more important to stress, however, are the implications of the analysis for the problem of employment. Focus on creation of technology rather than on embodiment of a given technique suggests in fact a different approach which has already brought about a shift of attention from labour inputs to human resources in the analysis of the process of innovation.

Focusing on employment implies referring to a productive capacity that embodies a given technology and that is characterized by certain labour inputs. Different levels of employment are then associated with different ways of working of the economy which are the expression of different (already established) technologies. However, when the attention is directed to the very process through which technological and productive transformations come about and acquire a given shape, focus on employment (that is, on the result of a process that must instead be first analysed in terms of viability and actual development) is misplaced. Conversely, the role played by human resources in the process of modification of the environment interpreted as a source of technology must be stressed.

Employment, in the sense of jobs provided in any period or over a given horizon, is then no longer to be considered as an objective of economic policy. It will in fact be the natural outcome of the way in which the process of innovation actually shapes the new productive capacity of the economy. The true target of economic policy becomes the viability of the process of innovation; and this, as we have seen, essentially depends on the intensity and the pace of the process of learning which results in new skills and qualifications of human resources. The intermediate targets (ΔM and Q, in the model) must be chosen accordingly.

Bibliography

Amendola, M. 1972: Modello Neo-Austriaco e transizione fra equilibri dinamici. *Note Economiche*, Nov.

—— 1983: A change of perspective in the analysis of the process of innovation. *Metroeconomica*, Oct.

—— 1984a: Towards a dynamic analysis of the traverse. *Eastern Economic Journal*, Apr–June.

—— 1984b: Productive transformations and economic theory. *Quarterly Review*, Banca Nazionale del Lavoro, Dec.

Amendola, M. and Gaffard, J. L. 1986: Technology as an environment: a suggested interpretation. *Economie Appliquée*, no. 3.

Benassy, J. P. 1975: Neo-Keynesian disequilibrium theory in a monetary economy. *Review of Economy Studies*, Oct.

—— 1976: The disequilibrium approach to monopolistic price setting and general monopolistic equilibrium. *Review of Economic Studies*, 43.

Böhm-Bawerk, E. 1889: *Positive Theory of Capital*. Huncke and Snnolz 1959.

Brunner, K. and Meltzer, A. 1971: The use of money in the theory of an exchange economy. *American Economic Review*, Dec.

Bruno, S. 1984: Heterogeneous labour, employment and distribution: a micro-macro theoretical framework for the analysis of segmented labour markets. *Economia e Lavoro*, no. 2.

Colombo, U. 1980: A viewpoint on innovation and the chemical industry. *Research Policy*, July.

Dosi, G. 1982: Technological paradigms and technological trajectories. A suggested interpretation of the determinants and directions of technical change. In C. Freeman (ed.), *Long Waves in the World Economy*, London: Butterworths, 1983.

—— 1984: *Technical Change and Industrial Tranformation*. London: Macmillan.

Drèze, J. 1975: Existence of an exchange equilibrium under price rigidities. *International Economic Review*, May.

109

Freeman, C., Clark, J. and Soete, L. 1982: *Unemployment and Technical Innovation*, London: Frances Pinter.

Friedman, M. 1971: A monetary theory of nominal income. *Journal of Political Economy*, 79, no. 2.

Georgescu-Roegen, N. 1965: Process in farming versus process in manufacturing: a problem of balanced development. In *Energy and Economic Myths*, Oxford: Pergamon, 1976.

—— 1974: Dynamic models and economic growth. In *Energy and Economic Myths*, Oxford: Pergamon, 1976.

Grandmont, J. M. and Laroque, G. 1976: On Keynesian temporary equilibrium. *Review of Economic Studies*, 43.

Hahn, F. 1952: Expectations and equilibrium. *Economic Journal*, Dec.

—— 1973: On the Foundations of Monetary Theory. In M. Parkin and A. R. Nobay (eds), *Essays in Modern Economics*, London, Longman Group, reprinted in *Equilibrium and Macroeconomics*, Oxford: Basil Blackwell, 1984.

—— 1974: On the notion of equilibrium in economics, inaugural lecture, Cambridge University, Cambridge: Cambridge University Press.

—— 1978: On non-Walrasian equilibrium. *Review of Economic Studies*, 45.

—— 1982: *Money and Inflation*. Oxford: Basil Blackwell.

Hahn, F. and Matthews, R. C. O. 1964: The theory of economic growth: a survey. *Economic Journal*, Dec.

Hayek, F. 1941: *Pure Theory of Capital*. London: Routledge and Kegan.

Hicks, J. R. 1932: *The Theory of Wages*, 2nd edn. London: Macmillan, 1963.

—— 1956: Methods of dynamic analysis. In *Collected Essays on Economic Theory*, vol. II, *Money, Interest and Wages*, Oxford: Basil Blackwell, 1982.

—— 1965: *Capital and Growth*: Oxford: Clarendon.

—— 1970: A neo-Austrian growth theory. *Economic Journal*, June.

—— 1973: *Capital and Time*. Oxford: Clarendon.

—— 1974a: *The Crisis in Keynesian Economics*. Oxford: Basil Blackwell.

—— 1974b: Industrialism. *International Affairs*, reprinted in *Economic Perspectives*, Oxford: Clarendon, 1977.

—— 1977: *Economic Perspectives*, Oxford: Clarendon.

—— 1979: *Causality in Economics*. Oxford: Basil Blackwell.

Jones, R. A. and Ostroy, J. M. 1984: Flexibility and Uncertainty. *Review of Economic Studies*, Jan.

Jonung, L. 1981: Ricardo on machinery and the present unemployment: an unpublished manuscript by K. Wicksell. *Economic Journal*, March.

Kamien, M. and Schwartz, N. L. 1982: *Market Structure and Innovation*, Cambridge Survey of Economic Literature. Cambridge: Cambridge University Press.

Keynes, J. M. 1930: A Treatise on Money. In *Collected Writings of J. M. Keynes*, vol. V, London: Macmillan, 1971.

Knight, F. H. 1921: *Risk, Uncertainty and Profit*. Boston: Houghton Mifflin.

Lindahl, E. 1930: Penning politikens medel, Engl. transl. In *Studies in the Theory of Money and Capital*, London: Allen and Unwin, 1939.

—— 1934: A note on the dynamic pricing problem, reprinted. In J. M. Keynes, *The General Theory and After. A Supplement. Selected Writings*, London: Macmillan, 1979.

—— 1939: *Studies in the Theory of Money and Capital*. London: Allen and Unwin.

Lowe, A. 1976: *The Path of Economic Growth*. Cambridge: Cambridge University Press.

Lucas, R. E. 1975: An equilibrium model of the business cycle. *Journal of Political Economy*, Dec.

Lundberg, E. 1937: *Studies in the Theory of Economic Expansion*, Reprints of economic classics. New York: A. M. Kelley, 1964.

Malinvaud, E. 1977: *The Theory of Unemployment Reconsidered*. Oxford: Basil Blackwell.

McCulloch, J. R. 1821: The opinions of Messrs. Say, Sismondi and Malthus on the effects of machinery and accumulation, stated and examined. *Edinburgh Review*, March.

Menger, C. 1871: *Principles of Economics*. Translation by J. Dingwall and B. Hoselitz, Glenex 1950.

Metcalfe, J. S. 1981: Impulse and diffusion in technical change. *Futures*.

Metcalfe, J. S. and Gibbons, M. 1983: On the economics of structural change and the evolution of technology. Paper presented at the 7th World Congress of the I.E.A. Madrid, Sept.

Morishima, M. 1969: *Theory of Economic Growth*. Oxford: Clarendon.

Myrdal, G. 1933: *Der Gleichgewichts begriff als Instrument der Geld theoretischen Analyse*, Wien, 1° ed. 1931. Engl. transl. In *Monetary Equilibrium*, W. Hodge and Co. Ltd, 1939.

Negishi, T. 1977: Existence of underemployment equilibrium. In *Equilibrium and Disequilibrium in Economic Theory*, Dordrecht: Reidel.

Nelson, R. R. and Winter, S. 1977: In search of useful theory of innovation. *Research Policy* 6.

——1978: Forces generating and limiting concentration under Schumpeterian competition. *Bell Journal of Economics*.

—— 1982: *An Evolutionary Theory of Economic Change*. Cambridge, MA: Belknap.

Pasinetti, L. L. 1965: A new theoretical approach to the problems of economic growth. In *Academiae Pontificiae Scientiarum Scripta Varia*, no. 28, Citta del Vaticano.

—— 1973: Vertical integration in economic analysis. *Metroeconomica*.

—— 1981: *Structural Change and Economic Growth*. Cambridge: Cambridge University Press.

Pavitt, K. 1984: Sectoral patterns of technical change: Towards a taxonomy and a theory. *Research Policy*.

Punzo, L. 1984: Accounting approach and multisectoral modelling. Institute for Advanced Studies, Research Memorandum 204, Wien.

Radner, R. 1968: Competitive equilibrium under uncertainty. *Econometrica*, Jan.

Ricardo, D. 1821: *The Principles of Political Economy and Taxation,* in *The Works and Correspondence of David Ricardo,* ed. Piero Sraffa, vol. I, Cambridge: Cambridge University Press, 1951.

Rosenberg, N. 1969: The direction of technological change: Inducement mechanisms and focussing devices. *Economic Development and Cultural Change,* reprint 1976.

—— 1976: *Perspectives on Technology.* Cambridge: Cambridge University Press.

—— 1982: *Inside the Black Box.* Cambridge: Cambridge University Press.

Sahal, D. 1981: Alternative conceptions of technology. *Research Policy,* 10.

Salter, W. E. G. 1960: *Productivity and Technical Change.* Cambridge: Cambridge University Press.

Schefold, B. 1976: Different forms of technical progress. *Economic Journal,* Dec.

—— 1979: Capital, growth and definitions of technical progress. *Kyklos,* vol. 32.

Schumpeter, J. 1934: *The Theory of Economic Development,* New York: Harvard University Press.

—— 1954: *History of Economic Analysis,* London: Allen and Unwin.

Smith, A. 1776: *An Inquiry into the Nature and Causes of the Wealth of Nations,* ed. R. Campbell and D. Skinner. Oxford: Clarendon, 1976.

Solow, R. M. 1957: Technical change and the aggregate production function. *Review of Economics and Statistics,* Aug.

—— 1962: Substitution and fixed proportions in the theory of capital, *Review of Economic Studies,* June.

—— 1967: The interest rate and transition between techniques. In *Socialism, Capitalism and Economic Growth: Essays Presented to M. Dobb,* ed. C. H. Feinstein. Cambridge: Cambridge University Press.

Spaventa, L. 1973: Notes on problems of transition between techniques. In J. Mirrless and N. H. Stern (eds) *Models of Economic Growth,* London: Macmillan.

Sraffa, P. 1951: *The Works and Correspondence of David Ricardo,* vol. I, Introduction. Cambridge: Cambridge University Press.

Tobin J. 1972: Inflation and unemployment. *American Economic Review,* May.

Wicksell, K. 1898: *Interest and Prices,* Engl. transl. of *Geldzins und Güterpreise.* London: Macmillan, 1936.

—— 1901: *Lectures on Political Economy,* vol. I. Engl. transl. London: Routledge and Sons, 1934.

Index

age structure of the processes of
  production, 63, 64, 75, 78, 91,
  95, 107
Amendola, M., 29, 32n, 46
approach, evolutionary (new), 3–4,
  10, 46; traditional, 1–3, 10, 46,
  59
appropriability, 7, 42
assets, financial, 40, 41, 46n, 53,
  83; liquid, 39–43, 51, 55, 75;
  real, 83; reserve, 40, 41, 46n, 51
availability of resources (inputs),
  2–3, 10, 106

barter economy, 32, 35, 41, 46, 53
Benassy, J. P., 81n
Bohm-Bawerk, E., 30, 35
Brunner, K., 60n

capital, fixed/circulating, 8–9; as a
  'fund', 19, 30, 41; as a stock of
  goods, 29–30, 56
capital goods, 21, 28–30, 53, 56
choice(s), financial, 102; flexible,
  16, 42–3; innovative, see
  innovative choice; real, 99;
  related sequence of, 34, 40, 57,
  73; routine, see routine choice
circular relations in production,
  28–9

circulation, financial, 41n, 51;
  industrial, 41n, 51
Clark, J., 5n
coefficients, technical, 2, 20, 58;
  vertically integrated, 25
Colombo, U., 12
comparison method, 1, 17–19
compensation, automatic, 18;
  problem (theory), 17–18
constraints, 47–9, 54, 86; financial,
  47, 49, 50–1, 55, 67, 72, 75, 83,
  96–7, 102; human resource, 15,
  47, 50–1, 55, 62–3, 67, 72, 75,
  77–8, 95, 99, 102; interaction
  between decisions and, 50–1; interaction between financial and human,
  50, 99; releasing of the, 51, 94, 96
consumers', degree of preference
  for the new output, 64, 72, 74,
  96; preference system, 6, 24, 38,
  55, 59, 64, 73–4, 94; utility
  function, 71, 73
consumption out of profits, 31, 67,
  73, 86
continuation analysis (theory), 37n

decisions, current production, 47,
  51, 54, 67, 69, 75; investment,
  47, 51, 54, 67, 72, 76, 78, 94
diffusion of innovation, 2, 3n

113